Real Shangnai Mathematics
Practice Book
4.2

This book is produced from independently certified FSC paper to ensure responsible forest management.
For more information visit: **www.harpercollins.co.uk/green**

William Collins' dream of knowledge for all began with the publication of his first book in 1819. A self-educated mill worker, he not only enriched millions of lives, but also founded a flourishing publishing house. Today, staying true to this spirit, Collins books are packed with inspiration, innovation and practical expertise. They place you at the centre of a world of possibility and give you exactly what you need to explore it.

Collins. Freedom to teach.

Collins
An imprint of HarperCollins*Publishers*
The News Building
1 London Bridge Street
London
SE1 9GF

Browse the complete Collins catalogue at
www.collins.co.uk

© Shanghai Schools (Pre-Schools) Curriculum Reform Commission 2007
© Shanghai Century Publishing Group Co., Ltd. 2018
© HarperCollinsPublishers Limited 2018

Published by arrangement with Shanghai Century Publishing Group Co., Ltd.

10 9 8 7 6 5 4

ISBN 978-0-00-826177-1

The educational materials in this book were compiled in accordance with the course curriculum produced by the Shanghai Schools (Pre-Schools) Curriculum Reform Commission and 'Maths Syllabus for Shanghai Schools (Trial Implementation)' for use in Primary 4 Second Term under the nine-year compulsory education system.

These educational materials were compiled by the head of Shanghai Normal University, and reviewed and approved for trial use by Shanghai Schools Educational Materials Review Board.

The writers for this book's educational materials are:

Editor-in-Chief: Jianhong Huang
Guest Writers: Jianhong Huang, Hui Tong, Peijing Xu

This volume's Practice Book was revised by: 'Primary School Maths Practice Book' Compilation Team

All rights reserved. No part of this publication may be reproduced, stored in a retrieval system, or transmitted in any form by any means, electronic, mechanical, photocopying, recording or otherwise, without the prior written permission of the Publisher or a licence permitting restricted copying in the United Kingdom issued by the Copyright Licensing Agency Ltd., Barnard's Inn, 86 Fetter Lane, London, EC4A 1EN.

British Library Cataloguing in Publication Data
A catalogue record for this publication is available from the British Library.

For the English edition:

Primary Publishing Director: Lee Newman
Primary Publishing Managers: Fiona McGlade, Lizzie Catford
Editorial Project Manager: Mike Appleton
Editorial Manager: Amanda Harman
Editorial Assistant: Holly Blood
Managing Translator: Huang Xingfeng
Translators: Chen Qingqing, Fan Lichen, Huang Chunhua, Liu Xuanru Lin Xumai, Lu Qingyi, Shi JiaminShi Zhiwei, Tang Xiaofen Yang Lili, Zhao Yang, Zhu Youqin
Lead Editor: Tanya Solomons
Copyeditor: Jan Schubert
Proofreaders: Deborah Oliver, Joan Miller
Cover artist: Amparo Barrera
Designer: Ken Vail Graphic Design
Production Controller: Sarah Burke
Printed and bound by CPI Group (UK) Ltd, Croydon, CR0 4YY

Photo acknowledgements
The publishers wish to thank the following for permission to reproduce photographs. Every effort has been made to trace copyright holders and to obtain their permission for the use of copyright materials. The publishers will gladly receive any information enabling them to rectify any error or omission at the fi rst opportunity.

(t = top, c = centre, b = bottom, r = right, l = left)

p28bl Shutterstock.com/ ConstantinosZ, p28br Shutterstock.com/ Egor Rodynchenko, p82 Shutterstock.com/ Vladimir Nenezic

All other images with permission from Shanghai Century Publishing Group.

Contents

Unit One: Revising and improving — 1

   1. Four arithmetic operations — 2
   2. Operation properties of whole numbers — 7
   3. Who is calculating cleverly? — 14
   4. Problem solving (1) — 17

Unit Two: Recognise decimal numbers and their addition and subtraction — 27

   1. Decimal numbers in real life — 28
   2. The meaning of decimal numbers — 29
   3. Comparing decimal numbers — 46
   4. Properties of decimal numbers — 48
   5. Practice exercise (1) — 50
   6. Multiplying and dividing decimals by multiples of 10 — 53
   7. Addition and subtraction of decimal numbers — 60
   8. The application of addition and subtraction of decimal numbers — 68
   9. Practice exercise (2) — 71

Unit Three: Statistics — 73

   1. Introduction to line graphs — 74
   2. Drawing a line graph — 79

Unit Four: Measure and geometry   83

    1. Perpendicular   84
    2. Parallel   90
    3. Practice exercise (3)   95

Unit Five: Consolidating and enhancing   97

    1. Problem solving (2)   98
    2. Decimal numbers and approximate numbers   103
    3. Perpendicular and parallel   108

Unit One: Revising and improving

A survey ship measures the depth of the sea in different places. It transmits sound waves from the sea surface to the sea floor.

After 8 seconds, it receives signals. What is the depth of the sea here, in metres? (Sound waves travel at about 1450 metres per second in seawater.)

The table below lists the sections in this unit.

After completing each section, assess your work.

(Use 😊 if you are satisfied with your progress or 😐 if you are not satisfied.)

Section	Self-assessment
1. Four arithmetic operations	
2. Operation properties of whole numbers	
3. Who is calculating cleverly?	
4. Problem solving (1)	

1 Revising and improving

1. Four arithmetic operations

Pupil Textbook pages 2–4   Level A

1. Calculate mentally and then write the answers.

75 + 25 × 4 =
25 − 25 ÷ 5 =
10 + 90 × 10 =
2 × 17 + 8 × 17 =
25 × 8 ÷ (25 × 8) =

420 ÷ (100 − 40) =
57 − 27 ÷ 3 =
8 × 8 ÷ 8 × 8 =
36 − 36 ÷ 36 + 36 =

2. True or false? Put a tick (✓) for 'true' or a cross (✗) for 'false' in the brackets.

a. 2630 − 867 + 133 = 2630 − (867 + 133) = 2630 − 1000 = 1630
( )

b. 78 + 22 × 5 = 100 × 5 = 500 ( )

c. 40 × 40 ÷ 20 = 40 × (40 ÷ 20) = 80 ( )

d. 95 × 99 + 99 = 95 × 100 = 9500 ( )

3. Work these out, showing the steps in your calculation. Use a shortcut strategy to calculate quickly.

675 − 999 + 325   4300 ÷ (5 × 43)   2012 − 65 × 12

630 ÷ (63 − 48)   167 + 133 × 10 + 90   (18 × 305 − 90) ÷ 15

Level B

Write an appropriate operation symbol in each ☐ to make the equations true. (You may add brackets if you need to.)

4 ☐ 4 ☐ 4 ☐ 4 = 0   4 ☐ 4 ☐ 4 ☐ 4 = 1
4 ☐ 4 ☐ 4 ☐ 4 = 2   4 ☐ 4 ☐ 4 ☐ 4 = 3
4 ☐ 4 ☐ 4 ☐ 4 = 4   4 ☐ 4 ☐ 4 ☐ 4 = 5

Pupil Textbook page 5

1. **Read and write these numbers.**

   a. 540 800   Read as: _____
      7 029 050   Read as: _____
      1 200 063 003 Read as: _____

   b. Nine million, six hundred and eight thousand and eight
      Write the number: _____

      Five hundred thousand, seven hundred
      Write the number: _____

      Three hundred and thirty million, five thousand and five
      Write the number: _____

      One billion, eight hundred and five million
      Write the number: _____

2. **Look at each number and match it to the same number in words.**

   3 030 303   Three million, three thousand and three

   3 003 003   Three million, thirty thousand and three

   3 033 000   Three million, three hundred and thirty

   3 000 330   Three million, thirty thousand, three hundred and three

   3 030 003   Three million, thirty-three thousand

3. **Use rounding to find the nearest number.**

   Round each number up to the nearest multiple of ten thousand.
   2 454 875 ≈ (   )   3 995 700 ≈ (   )

   Round each number up to the nearest multiple of 100 million.
   540 473 007 ≈ (   )   999 889 988 ≈ (   )

1 Revising and improving

Level B

1. Make up at least two seven-digit numbers with these seven digits: 0, 0, 0, 0, 0, 8 and 8.

2. A number that is rounded to the nearest multiple of ten thousand equals 450 000. The greatest number it could be is (   ). The smallest number it could be is (   ).

3. A five-digit number $a \approx 50\,000$ (the number is rounded to the nearest multiple of ten thousand). Another five-digit number $b \approx 48\,000$ (the number is rounded to the nearest multiple of a thousand). Which of the following statements is correct? (   ) (Write the letter of the correct answer in the brackets.)

    A. a must be greater than b.
    B. a and b cannot be equal.
    C. a may be less than b.

Pupil Textbook page 6 Level **A**

1. Work out the answers mentally and write them in the brackets.

4 × 24 − 8 ÷ 4 − 3 = (   )   (4 × 24 − 8) ÷ 4 − 3 = (   )

4 × (24 − 8 ÷ 4 − 3 = (   )   4 × [(24 − 8) ÷ 4 − 3] = (   )

4 × [24 − 8 ÷ (4 − 3)] = (   )   4 × (24 − 8 ÷ 4) − 3 = (   )

2. True or false? Put a tick (✓) for 'true' or a cross (✗) for 'false' in the brackets.

  a. 325 − 25 × 6 = 300 × 6 = 1800   (   )
  b. 54 × 36 + 36 × 46 = (54 + 46) × 36   (   )
  c. 125 × 8 ÷ 25 × 4 = (125 × 8) ÷ (25 × 4) = 10   (   )
  d. 8 × (11 × 9) = 8 × 11 × 8 × 9   (   )

3. Work these out, showing the steps in your calculation. Use a shortcut strategy to calculate quickly when possible.

544 − (233 + 244)   544 − 233 + 244

(585 + 415) ÷ 125 × 8   (585 + 415) ÷ (125 × 8)

4. Multiply these numbers. Use a shortcut strategy to calculate quickly.

  a. (61 + 25) × 4   **b.** 61 × 25 × 4

  c. 888 × 125   **d.** 37 × 25

1 Revising and improving

Level B

1. First work out the answers. Then try to find the rule behind the calculations. Apply the rule and give another example.

14 × 14 − 13 × 15   19 × 19 − 18 × 20   96 × 96 − 95 × 97

For example:

2. Use a shortcut strategy to calculate quickly: 144 × 45 + 90 × 28

3. Add brackets to make the equations true.

 a. 360 − 240 ÷ 12 + 8 = 6

 b. 360 − 240 ÷ 12 + 8 = 18

 c. 30 − 180 ÷ 5 × 12 = 792

2. Operation properties of whole numbers

Pupil Textbook pages 7–8 Level A

1. Use the properties of subtraction to write the appropriate operation symbol in each ◯ and write the correct number in each ☐.

   a. 862 − 293 − 61 = 862 ◯ (☐ ◯ ☐)

   b. 465 − (65 + 249) = 465 ◯ ☐ ◯ ☐

   c. 713 − (213 + 198) = ☐ ◯ ☐ ◯ ☐

2. True or false? Put a tick (✓) for 'true' or a cross (✗) for 'false' in the brackets.

   a. 345 − 126 + 174 = 345 − (126 + 174) = 345 − 300 = 45   ( )

   b. 1264 − (264 + 158) = 1264 − 264 + 158 = 1000 + 158 = 1158 ( )

   c. 372 − 125 − 172 = 372 − 172 − 125 = 200 − 125 = 75   ( )

3. Use the properties of subtraction to calculate the answers.

   124 − 73 − 27   587 − (87 + 179)

   1846 − 620 − 380   651 − (265 + 51)

   978 − 222 − 178   2365 − 365 − 190 − 810

1 Revising and improving

4. Read each question carefully and work out the answer.

  a. The tourist bus leaves Nanjing, going to Shanghai along the Shanghai–Nanjing Expressway. The Shanghai–Nanjing Expressway is 274 kilometres long. The bus travels 92 kilometres in the first hour and 104 kilometres in the second hour. How much further does the bus have to travel to reach Shanghai?

  b. Happy Primary School did some tree planting. They bought a total of 372 seedlings. They planted 138 trees on the first day and 172 trees on the second day. How many trees are left to plant? (Can you answer using two different methods?)

Level B

1. Subtract the numbers and write the answer in the brackets.

2000 − 22 − 24 − 36 − 38 − 45 − 55 = (   )

2. Fill in each ◯ with a '+' or '−' to make the equation true.

10 ◯ 20 ◯ 30 ◯ 40 ◯ 50 ◯ 60 ◯ 70 ◯ 80 = 0

Pupil Textbook pages 9–10

1. Use the properties of division to write the appropriate operation symbol in each ◯ and write the correct number in each ☐.

   a. 630 ÷ (5 × 63) = 630 ◯ ☐ ◯ ☐
   b. 90 000 ÷ 4 ÷ ☐ = ☐ ◯ (☐ ◯ 25)
   c. 12 000 ÷ ☐ ◯ 125 = ☐ ◯ (8 ◯ ☐)

2. True or false? Put a tick (✓) for 'true' or a cross (✗) for 'false' in the brackets.

   a. 2000 ÷ 125 ÷ 8 = 2000 ÷ (125 × 8)   ( )
   b. 300 ÷ 25 × 4 = 300 ÷ (25 × 4) = 300 ÷ 100 = 4   ( )
   c. 1500 ÷ 25 ÷ 15 = 1500 ÷ 15 ÷ 25 = 100 ÷ 25 = 4   ( )

3. Write <, = or > in each ◯ to make the sentence true, without working out the answers.

   848 ÷ 4 ÷ 2 ◯ 848 ÷ 8   96 − (54 + 26) ◯ 96 − 26 − 54
   36 ÷ 12 ÷ 3 ◯ 36 ÷ 4   360 ÷ 4 ÷ 9 ◯ 360 ÷ 2 ÷ 18
   720 ÷ (12 × 3) ◯ 720 ÷ 12 × 3
   3300 ÷ (33 + 33 + 33 + 33) ◯ 3300 ÷ 33 ÷ 4

4. Work these out, showing the steps in your calculation. Use a shortcut strategy to calculate quickly when possible.

   20 000 ÷ 125 ÷ 8   5600 ÷ (25 × 56)   1200 ÷ 25 ÷ 4

   10 400 ÷ 26 ÷ 4   7800 ÷ (39 × 40)   9595 ÷ (5 × 19)

1 Revising and improving

5. Read each question carefully and work out the answer.

　a. 12 March is tree planting day. Dylan and his classmates are divided into 4 groups to plant seedlings. Each team plants 15 seedlings. If the class bought all the seedlings with £300, how much did each seedling cost?

　b. The school library will donate 360 books to charity. Using Alex and Dylan's packing method, how many boxes can these books be packed into? (Try to answer this in two different ways.)

> Tie up the books with 12 books in each bundle.

> 6 bundles can be packed in a box.

Level B

1. Fill in each ☐ to make the equation true. How many different answers can you find? Write the answer on the line.

180 ÷ ☐ ÷ ☐ = 6

_____

2. Use a shortcut strategy to calculate quickly: 333 × 888 ÷ 999

Pupil Textbook pages 11–12 Level **A**

1. True or false? Put a tick (✓) for 'true' or a cross (✗) for 'false' in the brackets.

    a. 420 ÷ 60 = (420 ÷ 10) ÷ (60 × 10)   ( )

    b. 9300 ÷ 30 = (9300 ÷ 100) ÷ (30 ÷ 10)   ( )

    c. 45 ÷ 9 = (45 + 3) ÷ (9 + 3)   ( )

    d. 100 ÷ 25 × 5 = 100 ÷ (25 ÷ 5)   ( )

2. Work from top to bottom. The answers can be worked out by looking at the quotient from the first number sentence. Use this to help you write the correct answers in the brackets.

54 ÷ 9 = 6   450 ÷ 15 = 30

540 ÷ 90 = ( )   4500 ÷ 150 = ( )

5400 ÷ 900 = ( )   4500 ÷ 15 = ( )

54 000 ÷ 9000 = ( )   45 000 ÷ 150 = ( )

3. Read these problems and fill in the answers.

    a. When one number is divided by another number, the quotient is 102. If the dividend and the divisor are both divided by 6, the quotient is ( ).

    b. If the dividend is multiplied by 8, and the quotient remains the same, the divisor should be ( ) if there is no remainder.

    c. Write an operation symbol in each ◯ and put the correct number in each ☐.

       (60 ◯ ☐) ÷ (30 ◯ ☐) = 2

    d. 5600 ÷ 140 = 560 ÷ ( ) = ( ) ÷ 70 = ( ) ÷ ( )

1 Revising and improving

4. Look at the example in the shaded box. Then use the property of unchanged quotients to divide the other numbers.

$640 \div 40$
$= (640 \div 10) \div (40 \div 10)$
$= 64 \div 4$
$= 16$

$9100 \div 700$
$= (\quad) \div (\quad)$
$=$
$=$

$3600 \div 20$
$= (\quad) \div (\quad)$
$=$
$=$

$7000 \div 140$
$= (\quad) \div (\quad)$
$=$
$=$

Level B

1. How long will Emma take to finish 30 mental arithmetic exercises? Use her statement to find the answer.

I can do 10 mental arithmetic exercises in 30 seconds.

2. Calculate first. Think carefully. What do you find?

$9 \div 2 = (\quad)$ r $(\quad)$

$90 \div 20 = (\quad)$ r $(\quad)$

$900 \div 200 = (\quad)$ r $(\quad)$

$9000 \div 2000 = (\quad)$ r $(\quad)$

What I have found is: _____

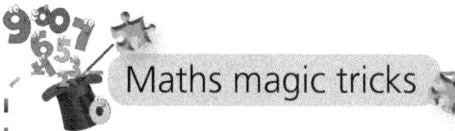 Maths magic tricks

Number guessing game

Poppy said to Dylan 'I can guess any three-digit number you think of. Dylan did not believe her. Poppy said 'Take out a piece of paper, pen or calculator and try it.'
Poppy asked Dylan do the following:

a. Think of any three-digit number, such as: 417.
b. Write this three-digit number twice, or input it into the calculator twice, to get a six-digit number, such as: 417417.

Then Poppy said, 'Your six-digit number must be a multiple of 7! It might be a multiple of 11! If I'm right, it should also be a multiple of 13!'

Dylan wondered whether every six-digit number derived from any three-digit number had such extraordinary results.

Poppy went on, 'Divide your six-digit number by 7, 11 and then 13 and see what happens!'

Dylan was surprised to find that the result was the three-digit number he first thought of!

Think carefully and try to work out the reason behind the answer. (You can also discuss this with your family and friends. Find out the answer later in this book.)

1 Revising and improving

3. Who is calculating cleverly?

Pupil Textbook page 13 — Level **A**

1. Work out these divisions, showing the steps in your calculation. Use a shortcut strategy to calculate quickly when possible.

   8000 ÷ 32   47 000 ÷ 94   53 000 ÷ 125

   26 000 ÷ 4 ÷ 25   5800 ÷ 25 ÷ 29   8100 ÷ (27 × 4)

2. Do a column division using the property of the unchanged quotient.

   2550 ÷ 150 =   28 400 ÷ 710 =

3. Multiple choice – write the letter of the correct answer in the brackets.

   a. 43 200 ÷ 40 = (   )

   A. 180   **B.** 18   **C.** 108   **D.** 1080

   b. The result of 72 ÷ 24 is the same as that of number sentence (   ).

   A. (72 × 2) ÷ (24 ÷ 2)   **B.** (72 ÷ 24) × (24 × 4)

   C. (72 + 72) ÷ (24 + 24 + 24)   **D.** (72 × 3) ÷ (24 × 3)

   c. The correct calculation procedure is (   ).

   A. 375 ÷ 25 = 375 × 4 ÷ 25 × 4 = 15

   B. 9600 ÷ (24 ÷ 8) = 9600 ÷ 24 × 8 = 32

   C. 125 × 8 ÷ 25 × 4 = 1000 ÷ 100 = 10

   D. 3200 ÷ (25 × 16) = 3200 ÷ 16 ÷ 25 = 200 ÷ 25 = 8

4. **Read each question carefully and work out the answer.**

   a. 240 pupils from a primary school go on a trip to the Science Museum. The pupils are divided into 6 equal teams and each team is divided into 4 groups. How many pupils are there in each group?

   b. There were 15 pupils in the art group at the youth club. They created 420 pictures in 7 months. How many pictures were created by each pupil in a month?

1 Revising and improving

Level B

1. Use the property of unchanged quotients to solve this problem: the quotient of 20 000 000 000 ÷ 12 500 000 is (　　).

2. Look at the patterns first and then fill in the missing numbers.

   a. 1 × 8 + 1 = 9
   12 × 8 + 2 = 98
   123 × 8 + 3 = 987
   1234 × 8 + 4 = (　　　　)
   12 345 × 8 + 5 + (　　　　)
   (　　　) × (　　　) + (　　　) = (　　　　)

   b. 11 × 11 = 121
   111 × 111 = 12 321
   1111 × 1111 = 1 234 321
   11 111 × 11 111 = (　　　　　)
   (　　　) × (　　　) = 1 234 567 654 321

   Can you find another set of equations that follow this kind of rule?

4. Problem solving (1)

Pupil Textbook page 14

Level A

1. Work these out, showing the steps in your calculation. Use a shortcut strategy to calculate quickly when possible.

   1923 − 456 − 544   152 × 12 + 48 ÷ 12   789 × 36 − 489 × 36

   22 000 ÷ 55   101 × 87   265 × 99 + 265

2. Read each question carefully and work out the answer.

   a. A school uniform coat costs £124 and a pair of trousers costs £76. There are 42 pupils in Class 1 of Year 4. If everyone buys a coat and a pair of trousers, what will the total cost be?

   b. Pupils in Year 4 were divided into two teams to plant trees. There were 38 pupils in each team. Team A planted 190 trees and Team B planted 266 trees.
   Which team planted more trees per person?
   How many more trees?

1 Revising and improving

c. The total distance of a road cycling race is 121 kilometres. A contestant has ridden 43 kilometres, and he expects to take 2 hours to reach the finishing line. If he cycles at a constant speed, what is his speed over the rest of the race?

d. Poppy and Emma went to the cinema from school. They walked along the same road and both started at the same time. Poppy walked 60 metres per minute; Emma walked 55 metres per minute. After 12 minutes, Poppy arrived at the cinema. How many metres away from the cinema was Emma at this time?

Level B

Work out the missing numbers and write them in the brackets.

a. 3600 ÷ 24 + 1200 ÷ 24 = (   ) ÷ 24

b. 25 × 6 + 25 × 7 + 25 × 8 + 25 × 9 − 25 × 10 = 25 × (   )

c. If 97 × 102 − 97 × ☆ − 97 × △ = 97 × 99, then ☆ + △ = (   )

Pupil Textbook page 15 Level A

1. **Work out these multiplications and divisions, showing the steps in your calculation. Use a shortcut strategy to calculate quickly when possible.**

$3900 \div (5 \times 39)$ $125 \times 888 \times 6$ $(205 \times 24 - 120) \div 25 \times 4$

2. **Write number sentences for the line segment diagrams below.**

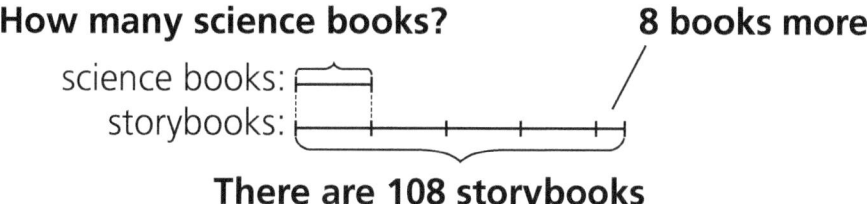

How many science books?
8 books more
science books:
storybooks:
There are 108 storybooks

Number sentence: ____________________

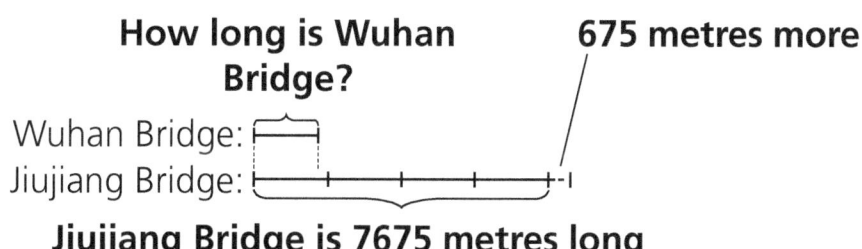

How long is Wuhan Bridge?
675 metres more
Wuhan Bridge:
Jiujiang Bridge:
Jiujiang Bridge is 7675 metres long

Number sentence: ____________________

3. **Multiple choice – write the letter of the correct answer in the brackets.**

   a. Poppy's weekly physical exercise time is 360 minutes, which is 30 minutes less than 3 times her mother's exercise time. How many minutes per week does her mother exercise? The correct number sentence is (   ).

   A. $360 \times 3 - 30$ **B.** $360 \div 3 - 30$

   C. $(360 - 30) \div 3$ **D.** $(360 + 30) \div 3$

1 Revising and improving

b. There are two piles of chess pieces. There are 87 pieces in the first pile. If 3 pieces are added to the second pile, the number of pieces in the first pile is 3 times the number in the second pile. So how many pieces are there in the second pile? The correct number sentence is (   ).

   A. 87 ÷ 3 B. 87 ÷ 3 + 3
   C. 87 ÷ 3 – 3 D. 87 × 3 – 3

c. Look at the tree diagram. What is the correct number sentence? (   ).

   A. 85 + 15 × 5 B. (85 + 15) × 5
   C. (85 – 15) ÷ 5 D. 85 – 15 ÷ 5

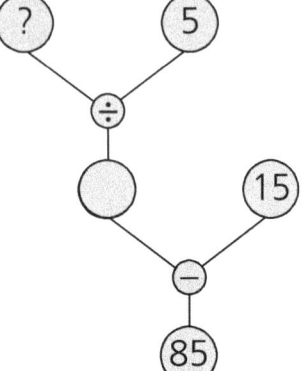

4. Read each question carefully and work out the answer.

a. A community's green space has an area of 12 000 square metres this year. This is 3000 square metres more than 2 times the area it was last year. How many square metres were there last year?

b. i. The mass of an elephant is 3 tonnes. The mass of a whale is 9 tonnes more than 37 times the mass of an elephant. What is the mass in tonnes of a whale?

ii. The mass of a whale is 120 tonnes. This is 9 tonnes more than 37 times of the mass of an elephant. What is the mass in tonnes of an elephant?

iii. During tree planting, Year 4 pupils planted 198 cypress trees. The number of pine trees they planted was 76 less than 4 times the number of cypress trees. How many trees were planted by the Year 4 pupils?

1 Revising and improving

 Level B

1. Work these out, showing the steps in your calculation.

   1200 ÷ 40 + 1800 ÷ 90   174 ÷ 3 + 276 ÷ 3

2. Read Emma and Dylan's speech bubbles. Then answer the question: How much money did Emma have at the start?

   When I give you £5, you have 5 times as much money as I do.

   I have £140 pounds. How much money did Emma have at the start?

Pupil Textbook pages 16–17

Level **A**

1. First finish each tree diagram. Then write the combined number sentence for each diagram.

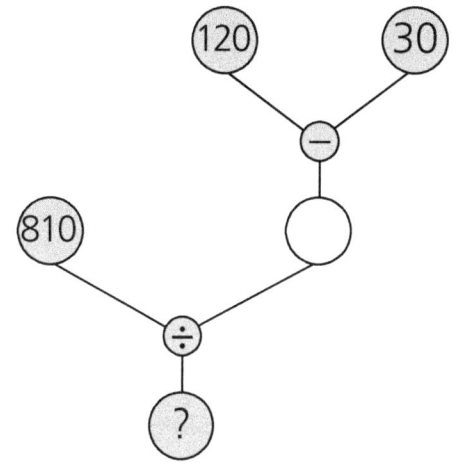

Combined number sentence:

_____

Combined number sentence:

_____

2. Multiple choice – write the letter of the correct answer in the brackets.

   I am thinking of two numbers, a and b.

   a. The value of a is 40 and b is 2 less than 3 times a. The correct number sentence to calculate b is (   ).

   A. 40 × 3 + 2 **B.** 40 × 3 − 2
   C. (40 − 2) ÷ 3 **D.** (40 − 2) ÷ 3

   b. The value of a is 40, which is 2 less than 3 times b. The correct number sentence to calculate b is (   ).

   A. 40 × 3 + 2 **B.** 40 × 3 − 2
   C. (40 + 2) ÷ 3 **D.** (40 − 2) ÷ 3

1 Revising and improving

3. Write the combined number sentence first. Then do the calculation and write the answer.

   a. A number that is 89 less than 50 times 24 is divided by 11. What is the quotient?

   b. A number is 9 less than 2 times 16. What is 3 times this number?

4. Read each question carefully and work out the answer.

   a. A dripping tap wastes about 43 litres of water a day. How many litres of water will 3 dripping taps, dripping at the same rate, waste in a month (31 days)?

   b. A rectangular playground was 50 metres long and 30 metres wide. The length and width are each increased by 8 metres. By how many square metres does the area increase? (Work out the answer in two ways.)

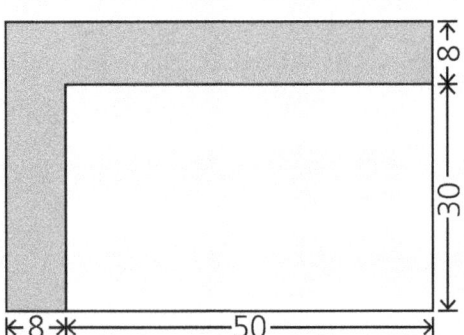

c. A supermarket buys 125 boxes of oranges and 125 boxes of apples. The mass of each box of oranges is 8 kilograms. The mass of each box of apples is 10 kilograms. How many more kilograms of apples are bought than oranges?

d. The school held a poetry contest. 32 pupils attended from Year 3. Twice as many Year 4 pupils attended, and the number of pupils attending from Year 5 was twice the total from Years 3 and 4. How many Year 5 pupils attended?

e. A primary school was collecting books for charity. The pupils of Year 3 donated 180 books and the number of books donated by Year 4 was 60 less than 2 times the number of books donated by Year 3. The number of books donated by Year 5 was 2 times that donated by Year 4. How many books were donated by Year 5?

f. The school bought some tables and chairs. It bought 120 sets the first time and it bought 145 sets the second time. The school spent £2625 more the second time than the first time. How much did it spend the first time?

1 Revising and improving

Level B

1. A survey ship measures the depth of the sea. It emits sound waves from the sea surface to the sea floor and receives signals in 8 seconds. Sound waves travel at about 1450 metres per second in the ocean. How deep is the sea?

2. The speed of the old bullet train can reach 250 km/h, which is 50 km/h more than 2 times the speed of an express train. The speed of the new bullet train is 3 times that of the express train. Calculate the speed of the new bullet train.

3. Two monkeys eat 1 kilogram of food in total every day. One panda eats as much food in a day as 72 monkeys eat in one day. One elephant eats 41 kilograms more than 2 times what one panda eats each day. How many kilograms of food does an elephant eat in a day?

Unit Two: Recognise decimal numbers and their addition and subtraction

1 kilogram of sugarcane can make 0.12 kilograms of sugar. How many kilograms of sugar can be made from 1 tonne of sugarcane? How about from 10 tonnes?

The table below lists the sections in this unit.

After completing each section, assess your work.

(Use 🙂 if you are satisfied with your progress or 😐 if you are not satisfied.)

Section	Self-assessment
1. Decimal numbers in real life	
2. The meaning of decimal numbers	
3. Comparing decimal numbers	
4. Properties of decimal numbers	
5. Practice exercise (1)	
6. Multiplying and dividing decimals by multiples of 10	
7. Addition and subtraction of decimal numbers	
8. The application of addition and subtraction of decimal numbers	
9. Practice exercise (2)	

2 Recognise decimal numbers and their addition and subtraction

1. Decimal numbers in real life

Pupil Textbook pages 19–20 *Level A*

1. Look at the prices and fill in the missing words and numbers.

   a. £24.50
   Read as:
   _____
   _____
   Expressed as:
   £ _____ and _____ pence

   b. £38.25
   Read as:
   _____
   _____
   Expressed as:
   £ _____ and _____ pence

   c. £40.08
   Read as:
   _____
   _____
   Expressed as:
   £ _____ and _____ pence

2. Use decimals to represent the prices of these goods.

Goods	Unit price	Decimal (units: £)
a pencil	85 pence	
a sack of rice	£12 and 9 pence	
a box of mixed fruit	£26 and 56 pence	

3. Find out the price (unit price per kilogram) of these vegetables.

£_____   £_____   £_____   £_____

Level B

Everyday measurements are not always whole numbers, but often include parts of units. We need to show these as decimal numbers. For example, Emma's height is 1.46 metres, a box of fruit has a mass of 6.25 kilograms, the highest temperature on one day is 26.5 °C, the capacity of a drinks carton is 0.98 litres… Can you think of some more examples?

2. The meaning of decimal numbers

Pupil Textbook pages 21–22 Level A

1. Use fractions and decimal numbers to represent the shaded parts.

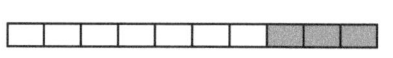

Fraction: _____
Decimal: _____

Fraction: _____
Decimal: _____

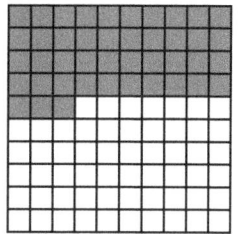

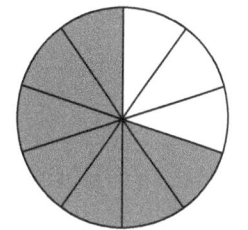

Fraction: _____
Decimal: _____

Fraction: _____
Decimal: _____

2. Write the correct decimal number in each ☐ on the number lines.

a.

b.
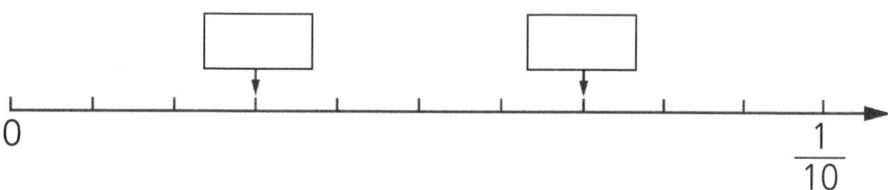

c.

2 Recognise decimal numbers and their addition and subtraction

3. Look at these prices and draw lines to match the amounts.

70 pence £$\frac{7}{100}$ £0.70

7 pence £$\frac{7}{10}$ £0.07

4. Fill in the missing numbers.

  a. Divide 1 into 100 equal pieces. Take 5 of them, and use decimals to represent this: (   ).

  b. 0.007 represents (   ) pieces if 1 is divided into (   ) equal pieces.

  c. 0.3 metres can represent (   ) pieces if (   ) is divided into (   ) equal pieces.

5. Write the corresponding fraction or decimal for each of these.

$\frac{4}{10}$ = _____ $\frac{4}{100}$ = _____ $\frac{4}{1000}$ = _____

0.05 = _____   0.6 = _____   0.007 = _____

Level B

Write the correct decimal number in each ☐.

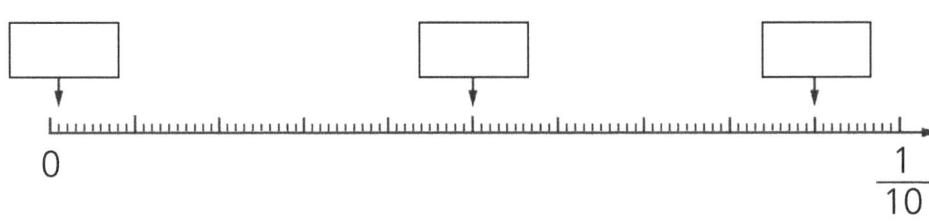

Pupil Textbook page 23

Level

1. Use decimal numbers to represent the shaded portions.

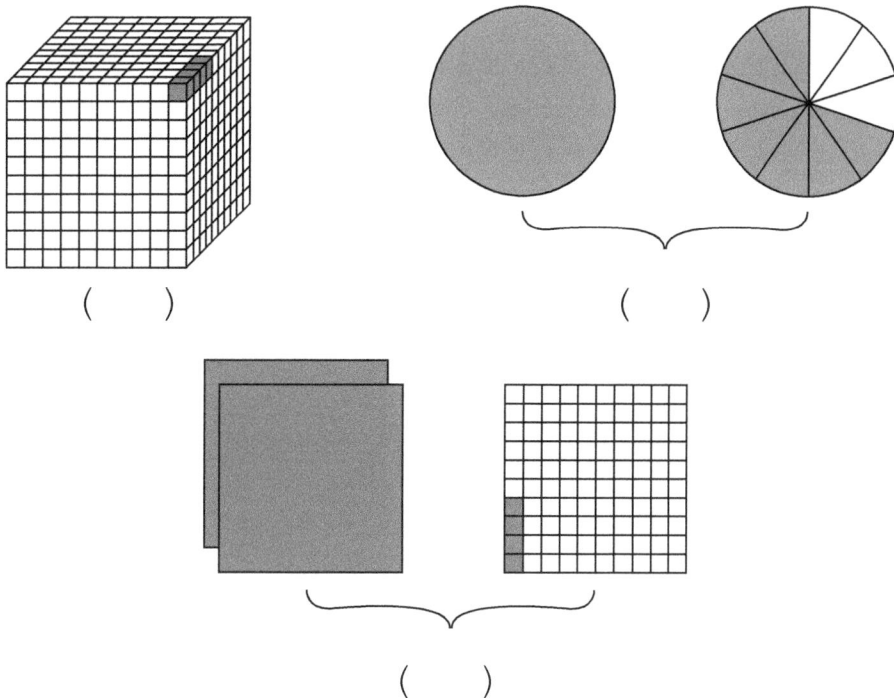

2. This square is 1 whole. Shade the square to represent 0.28 then fill in the missing numbers.

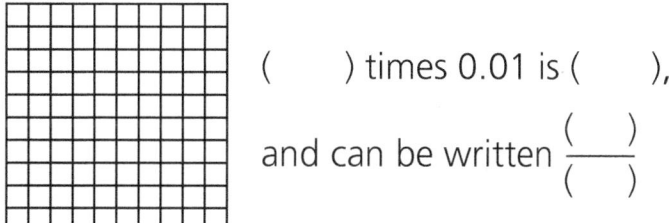

(　) times 0.01 is (　),

and can be written $\frac{(\ \ )}{(\ \ )}$

2. Recognise decimal numbers and their addition and subtraction

3. Write the answers in the spaces.

a. 7 times 0.1 is ____   17 times 0.1 is ____   117 times 0.1 is ____
   7 times 0.01 is ____   17 times 0.01 is ____   117 times 0.01 is ____
   7 times 0.001 is ____ 17 times 0.001 is ____ 117 times 0.001 is ____

b. ____ times 0.1 is 0.9   ____ times 0.01 is 0.69

c. 10 times 0.1 is ____   10 times 0.01 is ____
   100 times 0.1 is ____   100 times 0.01 is ____
   1000 times 0.1 is ____   1000 times 0.001 is ____

4. Write the correct decimal number in each ▢.

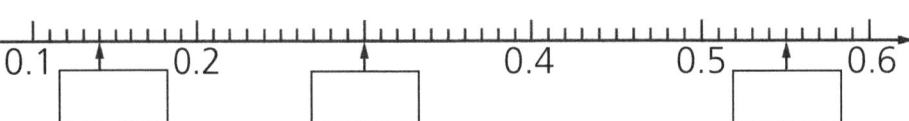

Level B

Write the correct fraction or decimal number in each box.

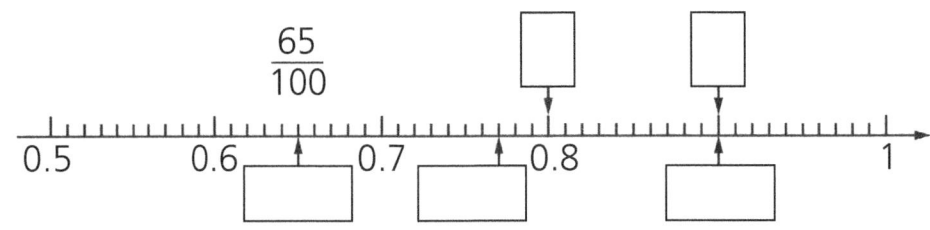

Pupil Textbook page 24   Level **A**

1. Write the correct decimal number in each ☐.

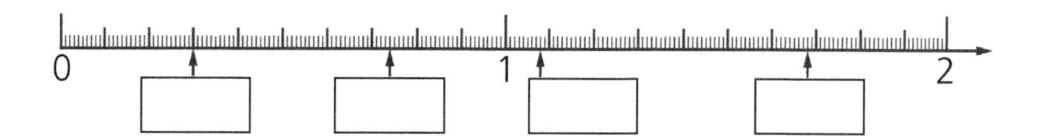

```
0 ─┬─────────┬──────────┬─────────┬──────── 2
   ↑         ↑      1   ↑         ↑
  ☐         ☐         ☐         ☐
```

2. Mark these decimal numbers on the number line. Then write the answers in the brackets below.

   72.18   72.6   73.02

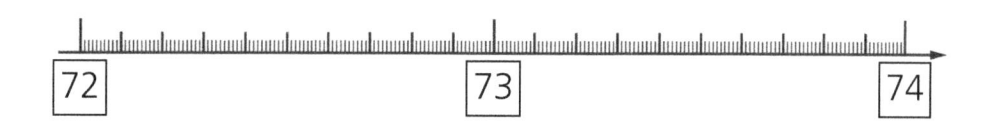

```
 ┌──┐              ┌──┐              ┌──┐
 │72│              │73│              │74│
 └──┘              └──┘              └──┘
```

   a. 72.18 = ____ × 10 + ____ × 1 + ____ × 0.1 + ____ × 0.01
   72.18 is composed of ( ) times 10, ( ) times 1, ( ) times 0.1 and ( ) times 0.01

   b. 72.6 = ____ × 10 + ____ × 1 + ____ × 0.1
   72.6 is composed of ( ) times 10, ( ) times 1 and ( ) times 0.1

   c. 73.02 is composed of ( ) times 10, ( ) times 1, ( ) times 0.1 and ( ) times 0.01
   73.02 = ____ × 10 + ____ × 1 + ____ × 0.1 + ____ × 0.01

3. Fill in the missing numbers.

   a. 18 times 0.001 is ( ); 300 times 0.1 is ( ); 300 times 0.01 is ( )

   b. 0.48 is composed of 48 times ( ); 2.05 is composed of 205 times ( )

   c. 107 times ( ) is 1.07; 170 times ( ) is 1.7

   d. 4 × 10 + 3 × 0.1 = ( )

   e. 70 times 1 and 7 times 0.01 is ( )

2. Recognise decimal numbers and their addition and subtraction

Level B

Think carefully and fill in the missing numbers.

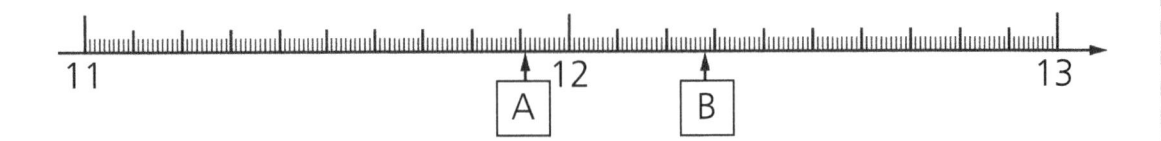

On the number line above, point A represents (   ). If it is increased by 0.01 times (   ), the result will be 12. Point B represents (   ). If it is decreased by 0.01 times (   ), the result will be 12.

Pupil Textbook pages 25–27

Level A

1. **Fill in the missing words and numbers.**

   a. A decimal number is 105.2309.

   The second digit to the right of the decimal point is (　), in the (　　　) place, expressed as (　　) times (　　).

   The fourth digit to the right of the decimal point is (　), in the (　　　) place, expressed as (　　) times (　　).

   The third digit to the left of the decimal point is (　), in the (　　　) place, expressed as (　　) times (　　).

   The first digit to the left of the decimal point is (　), in the (　　　) place, expressed as (　　) times (　　).

   b. 5 times 10 and 2 times 0.1 is (　　).

   c. (　　　) is composed of 3 times 100, 2 times 10, 7 times 1, 6 times 0.1 and 8 times 0.01.

   d. 20.903 is composed of (　) times 1, (　) times 0.1, (　) times 0.01 and (　) times 0.001.

   e. 160.85 is composed of (　) tens, (　) tenths, (　) hundredths.

   f. 24.38 is composed of (　) times 1 and (　) times 0.01.

   g. 4.7 is (　) times 0.1; 523 times 0.01 is (　).

2. **Write the meaning of the 6 in each of these numbers.**

   6 in 2.**6**5 means _____.

   6 in **6**0.25 means _____.

   6 in 0.25**6** means _____.

   6 in **6**00.52 means _____.

2 Recognise decimal numbers and their addition and subtraction

3. Write these numbers on the correct lines.

1.02   0.176   30.03   0.9   10.003   0.008   0.83   33.3

Pure decimals: _____
Mixed decimals: _____
Decimals with 1 decimal place: _____
Decimals with 2 decimal places: _____
Decimals with 3 decimal places: _____

Level B

1. Use the three digits 0, 5 and 8 and a decimal point to compose these numbers.
A decimal number with a digit in the tenths place:
(   )
A decimal number with a digit in the hundredths place:
(   )

2. Use any of the digits 0, 0, 1, 2 and 3, and a decimal point, to write the greatest pure decimal number (   ), and the smallest pure decimal number (   ).

3. A decimal number has 2 decimal places, and has the following characteristics:
- The whole number part is 1 less than the smallest three-digit number.
- The digit in the tenths place is larger than the digit in the hundredths place, and their sum is 10 while their difference is 2.

This decimal is (   ).

Can you write this mixed decimal?

Pupil Textbook page 28   Level **A**

1. Use decimal numbers to represent these amounts.

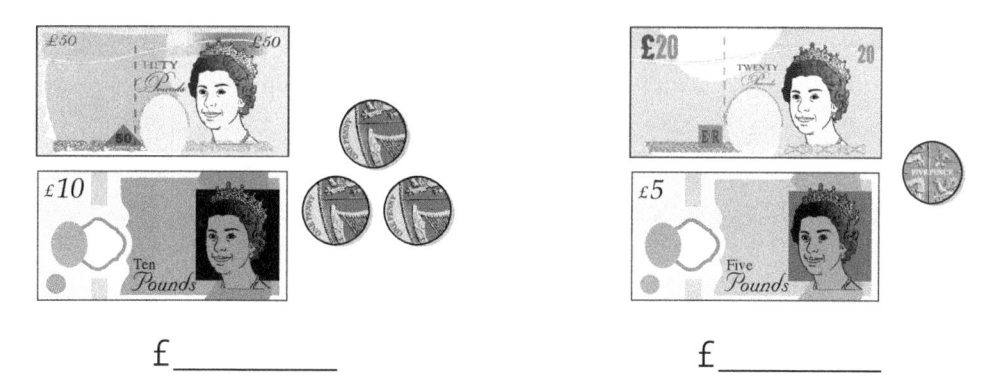

£_____   £_____

2. Write the correct numbers in the brackets.

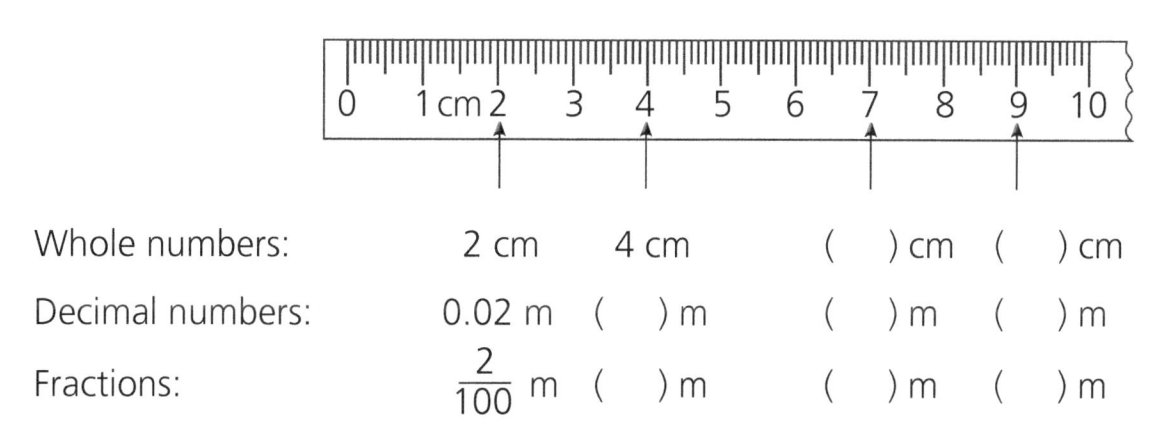

Whole numbers: 2 cm   4 cm   ( ) cm   ( ) cm
Decimal numbers: 0.02 m   ( ) m   ( ) m   ( ) m
Fractions: $\frac{2}{100}$ m   ( ) m   ( ) m   ( ) m

3. Look at these measurements and draw lines to connect those that match.

$\frac{3}{10}$ m $\frac{3}{10}$ cm   30 cm

$\frac{3}{100}$ m   0.3 m   3 cm

0.3 cm   0.03 m   3 mm

2. Recognise decimal numbers and their addition and subtraction

Level B

The length of this wire is about (   ) metres.

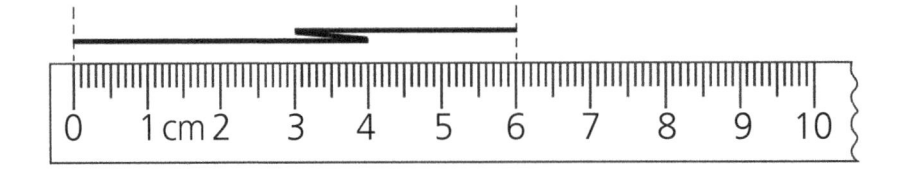

Magic maths

The secrets of the guessing game on page 11

Poppy asked Dylan to repeat the three digits in his head to get a six-digit number. This is, in fact, this three-digit number times 1001, and 1001 = 7 × 11 × 13. So you divide the six-digit number by 7, 11 and 13. The result is the number that Dylan was thinking of!

For example, Dylan thought of 417, and repeated it: 417 417

417 417 = 417 × 1001, but 1001 = 7 × 11 × 13

So, 417 417 ÷ 7 ÷ 11 ÷ 13 = 417

Pupil Textbook page 29　　　　　　　　　　　　　Level **A**

1. Measuring

   a. What is the length of this stick?

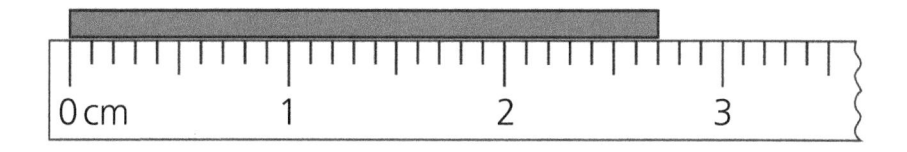

   The length of the stick is (　　) cm.

   b. What is the length of this stick?

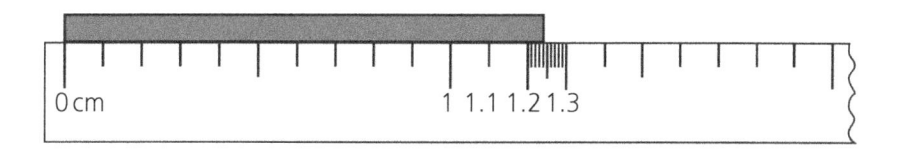

   The length of the stick is (　　) cm.

   c. What are the lengths of A and B?

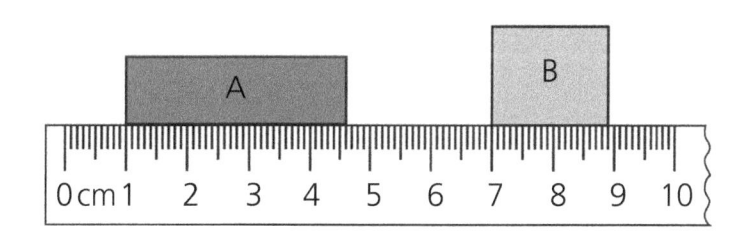

   The length of A is (　　) cm.　The length of B is (　　) cm.

2. The diagram shows how two set squares can be used to measure the diameter of the face of the small alarm clock.

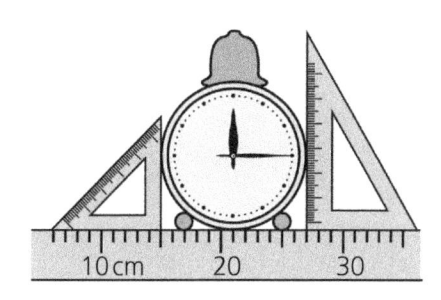

The diameter of the face of this small alarm clock is (　　) cm.

39

2 Recognise decimal numbers and their addition and subtraction

3. Practice

Measure some objects around you, such as this textbook, your desk, the teacher's desk. Write the measurements in the brackets.

a. The length of the maths book is ( ) cm and the width is ( ) cm.

b. The length of the desk is ( ) cm and the width is ( ) cm.

c. The length of the teacher's desk is ( ) cm and the width is ( ) cm.

Level B

Use a ruler and a set square to measure the length and diameter of the screw in the picture.

The length of the screw is ( ) cm.

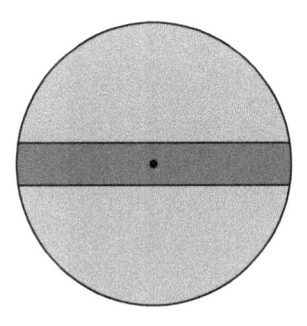

The diameter of the head of the screw is ( ) cm.

Pupil Textbook page 30

Level A

1. Use decimal numbers to represent the length of each object.

   a.

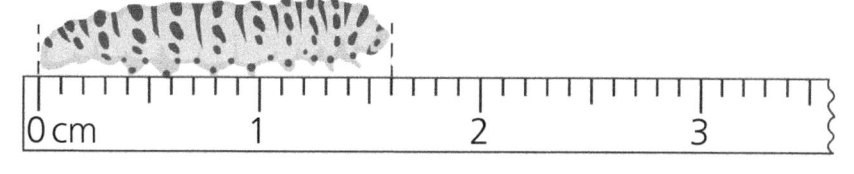

   The length of the caterpillar is (   ) cm.

   b.

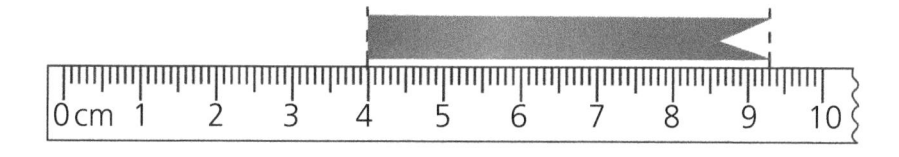

   The length of the ribbon is (   ) cm.

   c.

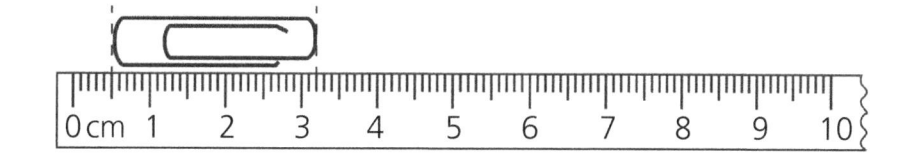

   The length of the paperclip is (   ) cm.

2. Write the correct fraction in each ◯ and write the correct decimal number in each ☐.

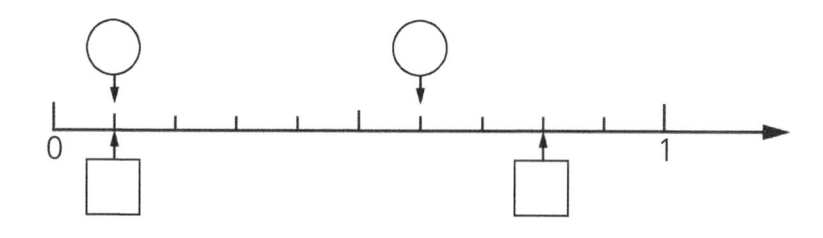

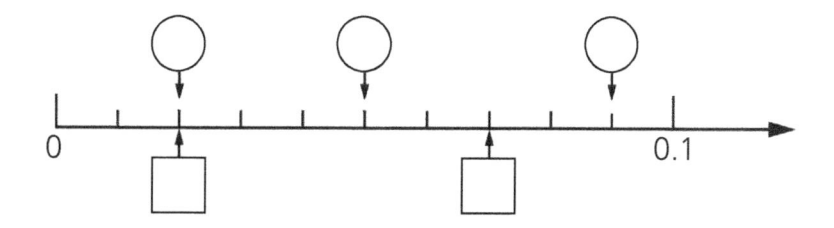

2 Recognise decimal numbers and their addition and subtraction

3. First mark the following decimal numbers on the number line. Then partition each decimal number. The first one has been done for you.

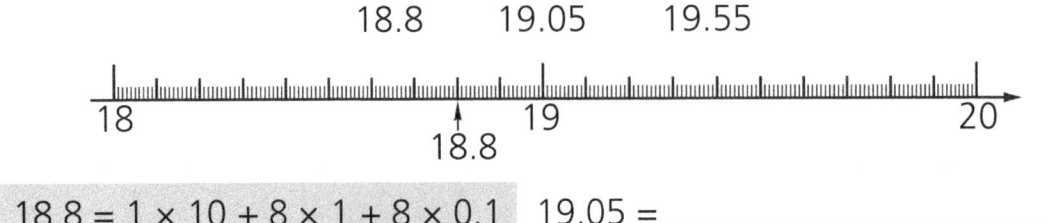

18.8   19.05   19.55

18 ———— 18.8 ———— 19 ———————— 20

18.8 = 1 × 10 + 8 × 1 + 8 × 0.1   19.05 = _____

19.55 = _____

4. Fill in the missing numbers.

  a. 3.25 is composed of ( ) times 1, ( ) times 0.1 and ( ) times 0.01.

  b. ( ) is composed of 4 times 0.1 and 8 times 0.001.

  c. 85 times 0.01 is ( ); 23 times 0.001 is ( ); 13 times 0.1 is ( ).

  d. Both the ones digit and thousandths digit of a number with 3 decimal places are 5, and all the other digits are 0; this number is ( ).

  e. (   ) is composed of three hundreds, eight ones and six hundredths.

  f. 8 in the decimal part of 85.768 is the (   ) digit, and is represented as ( ) times ( ); 8 in the whole number part of 85.768 is the ( ) digit, and is represented as ( ) times ( ).

  g. 7 centimetres represented as a fraction is ( ) metre.

  h. The second digit to the left of the decimal point is the ( ) digit, and the second digit to the right of the decimal point is the ( ) digit.

Level B

Use 2, 0, 5, 8 and a decimal point to make these numbers.

A pure decimal number: (   )
A mixed decimal: (   )
The largest pure decimal number: (   )
The smallest pure decimal number: (   )
The smallest decimal with 3 decimal places: (   )

Pupil Textbook pages 31–34

Level **A**

1. **Read these decimal numbers.**

   a. The Earth takes about 365.25 days to orbit the Sun.
      365.25 is read as: _____

   b. The Moon's diameter is about 3476.4 kilometres.
      3476.4 is read as: _____

   c. A table tennis ball weighs 2.71 grams.
      2.71 is read as: _____

2. **Write these as decimal numbers.**

   a. A big African beetle is fourteen point eight nine five cm long and has a mass of ninety-nine point seven nine grams.

      fourteen point eight nine five is written as: _____

      ninety-nine point seven nine is written as: _____

   b. The circumference of the Earth at the equator is approximately nine million, seven hundred and five point six nine kilometres.

      nine million, seven hundred and five point six nine is written as: _____

3. **Fill in the missing words and numbers.**

   a. A number is composed of five hundreds, two tenths and six hundredths. This number is read as: _____
      _____

   b. A number is composed of nine units, eight tenths and three hundredths. This number is read as: _____
      _____

   c. A baby panda has a mass of about 0.102 kilograms. An adult panda has a mass of about one hundred point five kilograms.
      0.102 is read as: _____
      one hundred point five is written as: _____

43

2. Recognise decimal numbers and their addition and subtraction

d. i. Read the numbers and fill in the missing words and numbers.
One kilowatt hour of electricity is needed to produce about 1.61 kilograms of steel; read as: _____ kg.
It takes a kilowatt hour of electricity to produce zero point four zero five kilograms of coal; written as: _____ kg.

ii. In the two decimal numbers in part i., the mixed decimal number is _____ and the pure decimal number is _____.

4. Complete the table of numbers written to 3 decimal places.

Read as	Write as
	3.001
fifty-two point zero eight zero	
	50.010
three point three zero zero	
	101.101
zero point two zero two	

5. Draw lines to match the numbers to the words.

606.06   sixty point double zero six

660.606   six hundred point double zero six

60.006   six hundred and six point zero six

606.6   six hundred and six point six

600.006   six hundred and sixty point six zero six

6. Choose five books you like and write the price of each book in the table. Read the price and say it.

Book					
Price (£)					

Level B

Use these cards to make numbers. Arrange the five cards into decimal numbers and read them. (Write two decimal numbers for each question.)

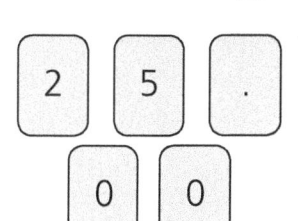

Cards: 2, 5, ., 0, 0

a. The decimal number with 2 decimal places where only one zero is said is: _____;

b. The decimal number with 1 decimal place where no zeros are said is: _____

2 Recognise decimal numbers and their addition and subtraction

3. Comparing decimal numbers

Pupil Textbook pages 35–36 Level **A**

1. Mark the decimal numbers on the number line and compare them.

   a. 11.35 and 12.53

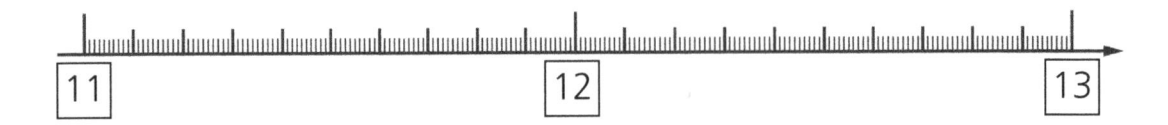

   |11| |12| |13|

   b. 1.65 and 1.09

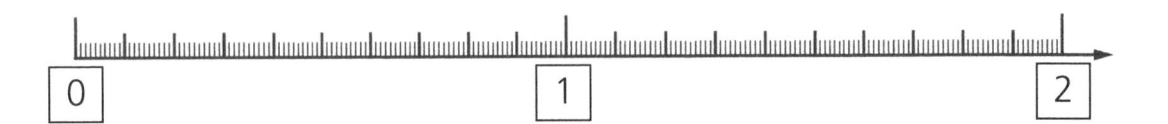

   |0| |1| |2|

2. a. Compare these numbers and write <, = or > in each ☐.

   3.01 ☐ 2.99   9.44 ☐ 9.46   15.60 ☐ 15.6
   3.567 ☐ 3.576   1 ☐ 0.999   10.10 ☐ 10.1

   b. Compare these numbers and write the correct numbers in the brackets.

   0.95 and 0.59   (   ) < (   )
   1.05 and 0.9985   (   ) < (   )
   1.19 m and 1.25 m   (   ) < (   )

3. Order these decimal numbers from smallest to greatest.

   a. 0.5, 0.51, 0.501, 0.511

   b. 4.56, 5.65, 4.565, 4.506

4.

Name	Time (seconds)
Emma	58.52
Poppy	58.17
Dylan	59.02
Alex	58.09

In the aircraft model competition, whoever's aircraft stayed in the sky the longest was the winner.

The above table records the flight times in the aircraft model competition. Can you put them in order of flight time?

_____

Level B

What numbers can go in the ☐?
(Write the answers on the line.)

6.43 > ☐.5

Numbers that can go in the ☐:

_____

0.214 < 0.2☐5

Numbers that can go in the ☐:

_____

2 Recognise decimal numbers and their addition and subtraction

4. Properties of decimal numbers

Pupil Textbook pages 37–40 Level A

1. Use the property of decimal numbers to simplify each of these decimal numbers.

5.40 = 11.050 = 2.700 = 10.800 =

0.200 = 0.0710 = 140.00 = 0.8030 =

2. Change each of these into a number with 3 decimal places without changing its value.

0.9 = 4.2500 = 10.5 =

13 = 58.4 = 0.80 =

3. True or false? Put a tick (✓) for 'true' or a cross (✗) for 'false' in the brackets.

a. The value of a decimal number will not change if zeros are inserted or removed from the end of the digits after the decimal point. ( )

b. 10.040 can be simplified to 10.4 ( )

c. 0.500 = 0.5 ( )

d. 0.018 is equal to 0.180 ( )

e. 100 thousandths is equal to one tenth ( )

f. Compare two decimal numbers: the more digits there are in the decimal part, the greater the value of the decimal number. ( )

4. Fill in the brackets.

a. 0.6 is (   ) times 0.1 and (   ) times 0.001.

b. Change 3.8 into a number with 3 decimal places without changing the value of the number. (   )

c. Compare 12.1 and 12.01: (   ) is greater.

d. 10.0400 can be simplified to (   ); 4.1600 can be simplified to (   ).

e. Change 1.7 into a number with 2 decimal places: (   )
Change 9 into a number with 3 decimal places: (   )

Level B

What numbers can go in ☐☐ ?
(Write the answers on the line.)

$$0.36 < 0.3\boxed{}\boxed{} < 0.37$$

Numbers that can go in ☐☐ :

_____

_____

2 Recognise decimal numbers and their addition and subtraction

5. Practice exercise (1)

Pupil Textbook pages 41–42 Level **A**

1. Shade and compare. (Write <, = or > in each ◯.)

I divide a square into 10 equal parts and shade one of the 10 parts.

I divide a square of the same size into 100 equal parts and shade ten of the 100 parts.

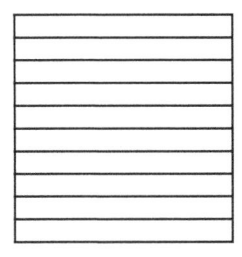

the coloured part shaded by Emma 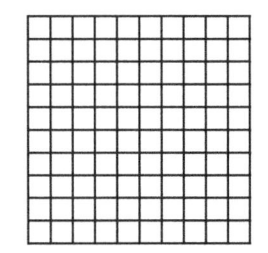   ◯   the coloured part shaded by Alex

2. Write a fraction in each ◯ and write a decimal number in each ☐.

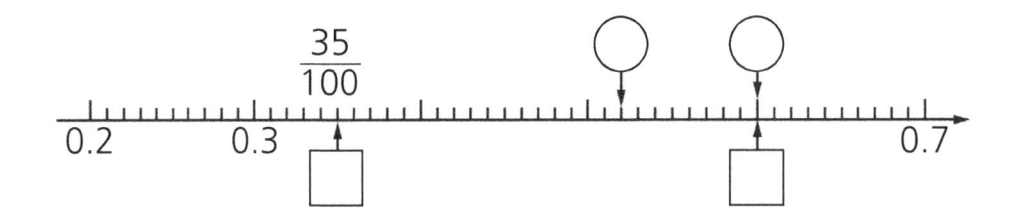

$\frac{35}{100}$

0.2   0.3   ☐   ◯   ◯   ☐   0.7

3. Look at these decimal numbers and fill in the missing numbers.

    a. Rewrite 0.6 as a decimal number with 2 decimal places: ____

    b. In 10.01, '1' on the left of the decimal point is in the ____ place, representing ____ times ____. '1' on the right of the decimal point is in ____ place, representing ____ times ____.

    c. 52.07 is made of ____ tens, ____ ones, ____ tenths and ____ hundredths. It can be expressed by the following formula:

    52.07 = ____ × 10 + ____ × 1 + ____ × 0.1 + ____ × 0.01

    d. 60.00 can be simplified to ____ by using the property of decimals numbers.

    If the size of the number is not changed, 0.06 can be rewritten as ____.

    e. 13 times $\frac{1}{100}$ can be expressed as a fraction: ____; or as a decimal number: ____

    f. 4 times 10 and 97 times $\frac{1}{100}$ make up the number ____.

4. Multiple choice – write the letter of the correct answer in the brackets.

    a. In the following numbers, write '0' at the end. Which number changes in size? (   )

    A. 5.1   **B.** 36   **C.** 20.7   **D.** 10.00

    b. Which number is equal to 5.07? (   )

    A. 5.007   **B.** 0.507   **C.** 5.070   **D.** 50.07

2　Recognise decimal numbers and their addition and subtraction

5. Which of these decimal numbers should be written in the ovals?

0.58　　50.8　　0.052　　8.25　　52.8　　25.08　　0.805　　820.5

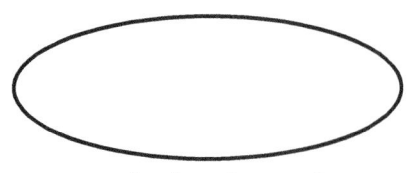

pure decimal numbers with 3 decimal places

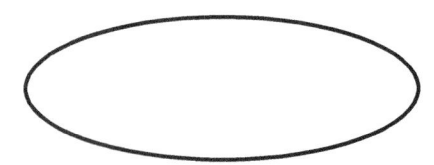

decimal numbers with 2 decimal places

6. Look at these comparisons and write a suitable number in each ☐.

3.056 < 3.05☐　　　　9.☐6 > 9.☐7　　　　3.4☐5 > 3.48

7. Dylan, Alex, Poppy and Emma have a long jump competition. The distances they jumped are:

Alex: 1.26 m;　Dylan: 1.20 m;　Poppy: 0.99 m;　Emma: 1.09 m

Put them in order, based on the distances.
First:　 (　　)　　　distance jumped: (　　　)
Second: (　　)　　　distance jumped: (　　　)
Third:　 (　　)　　　distance jumped: (　　　)
Forth:　 (　　)　　　distance jumped: (　　　)

Level B

Guess the number: it is a decimal number; this number is less than 3; the digit in the ones place is greater than 1; the digits in the ones place and tenths place are the same. What is this number?

_____

6. Multiplying and dividing decimals by multiples of 10

Pupil Textbook pages 43–47　　　　　　　Level A

1. Fill in the missing words and numbers.

   a. When a decimal number is multiplied by 10, the digits move (　　) place to the (　　).

   b. When the digits move (　) place to the right, the original number is (　　).

   c. When the digits of A move (　　) places to the left, the answer is equal to B. B is (　　) times greater than A.

   d. When the digits of a number are moved (　　) places to the left, then moved one place to the right, the result is 4.02. The original number is (　　).

2. Work out these multiplications and divisions.

   3.8 × 100 =　　　　0.2 ÷ 10 =　　　　2.006 × 10 =

   12.4 ÷ 100 =　　　15 ÷ 1000 =　　　1.003 × 100 =

   3.76 ÷ 100 =　　　100.2 × 100 =

3. Write × or ÷ in each ◯, and write the correct number in each ▢.

   0.53 ◯ ▢ = 53　　　　　40 ◯ ▢ = 0.04

   0.013 ◯ ▢ = 0.0013　　8.2 ◯ ▢ = 8200

   ▢ ÷ 10 000 = 0.0067　　▢ × 100 = 0.45

2. Recognise decimal numbers and their addition and subtraction

4. Multiple choice – write the letter of the correct answer in the brackets.

a. 30.15 (   ) is 3015; 30.15 (   ) is 3.015.

　　A. × 100　　　B. ÷ 100　　　C. ÷ 10

b. Write two zeros at the end of the decimal 3.5. Comparing the new number with the original number: (   ).

　　A. the original number is greater
　　B. the original number is smaller
　　C. the numbers are equal

c. First, move the digits of 1.717 two places to the left, and then move the digits three places to the right. Comparing the final number with the original number: (   ).

　　A. the original number is greater
　　B. the original number is smaller
　　C. the numbers are equal

5. Read the following problems and write the answers in the spaces below.

a. A clothing factory is making coats. Each coat needs 3.2 m of fabric. How many metres of fabric are needed to produce 1000 coats?

b. 1 kilogram of sugar cane can make 0.12 kilograms of sugar. How many kilograms of sugar can be made from 1 tonne of sugar cane make?

Level B

1. A boy wrote a decimal number, but he forgot the decimal point. As a result, he wrote 29 004. When the whole number is read aloud, only one zero is said. The number is: (　　　　). Moving the digits of the decimal number two places to the right, we get (　　　　).

2. There are three numbers, A, B and C. If we move the digits of A two places to the left and the digits of B three places to the right, the two numbers we obtain are equal to C. If C is 8.04, then A is (　　　) and B is (　　　).

2 Recognise decimal numbers and their addition and subtraction

Pupil Textbook pages 48–49 Level **A**

1. Calculate and fill in the missing numbers.

The price of the notebook:
£3.60 = (   ) pence

The length of the Nine-dragon Screen of the North Sea in Beijing:
25.86 m = (   ) cm

The capacity of an electric kettle:
2100 ml = (   ) l

The speed of sound waves in water is about 1450 metres per second:
1450 m = (   ) km

The capacity of a bottle of vegetable oil is 2.5 litres:
2.5 l = (   ) ml

The area of the desktop is about 4000 cm²:
4000 cm² = (   ) m²

2. Convert these units and write the answers in the brackets.

27 m = (   ) km   3.5 km = (   ) m   880 cm² = (   ) m²
700 g = (   ) kg   5 ml = (   ) l   £6.90 = (   ) pence
(   ) cm = 0.43 m   10 km 90 m = (   ) km
(   ) g = 20 kg 8 g

3. Write <, = or > in each ◯.

800 cm ◯ 0.8 m   £3 and 5 pence ◯ £3.50
7 l 45 ml ◯ 7.045 l   7.02 m² ◯ 7 m² 20 cm²

4. Work out the answers to these additions and subtractions and write them in the brackets.

$0.78\,l + 22\,ml = ($ $)\,ml$ $2.5\,kg = 725\,g = ($ $)\,g$

$3.09\,dm^2 - 45\,cm^2 = ($ $)\,cm^2$ $0.06\,km + 60\,m = ($ $)\,m$

$10.1\,m = ($ $)\,m + ($ $)\,cm$ $10.1\,m^2 = 10\,m^2\,($ $)\,cm^2$

5. Read each question carefully and work out the answer.

   a. Dylan walked 0.5 km in 10 minutes. How many metres did he walk per minute?

   b. 1 tonne of wheat can make 850 kg of wheat flour. How many kilograms of wheat flour can be made with one kilogram of wheat?

Level B

There are three sticks of different colours, of lengths $\frac{1}{2}$ m, 0.65 m and 55 cm. The red stick is longer than the blue stick but shorter than the green stick. What are the lengths of the red, blue and green sticks?

2 Recognise decimal numbers and their addition and subtraction

Pupil Textbook pages 49–50   Level **A**

1. Rewrite these numbers as numbers with units of 'thousand', 'million' or 'billion'.

328 190 = _____ thousands   507 200 000 = _____ billions

8002 = _____ thousands   14 600 000 = _____ millions

2. Rewrite these numbers in numerals.

10.9 thousands =   0.063 billions =
_____   _____

0.28 thousands =   500.7 millions =
_____   _____

3. Work out the answers and write them in the brackets.

   a. The distance between the Earth and the Moon is about 384 400 kilometres. 384 400 kilometres = (   ) thousand metres.

   b. In China's sixth census the population was 1 370 536 875 people. Rounding to the nearest billion, this is about (   ) billion people.

   c. Shanghai Pudong International Airport is one of the busiest in China. In 2013, there were 371 222 aircraft landings, a passenger throughput of 47 191.8 thousand people and a cargo throughput of 2914.8 thousand tonnes.

   371 222 landings = (   ) thousands

   2914.8 thousand tonnes = (   ) million kg

   47 191.8 thousand people = (   ) million people

4. Multiple choice – write the letter of the correct answer in the brackets.

   a. Change 7 to a number with 3 decimal places. (   )

      A. 0.007

      B. 7.000

      C. 0.700

      D. 0.070

   b. Which of these masses is the greatest? (   )

      A. 33600 g

      B. 36.36 kg

      C. 0.363 t

      D. 336 kg

Level B

A decimal number is formed from the six digits 2, 3, 4, 5, 6 and 8 and a decimal point. This decimal number rounds to approximately 5.8. What is the maximum this decimal number could be?

2 Recognise decimal numbers and their addition and subtraction

7. Addition and subtraction of decimal numbers

Pupil Textbook pages 51–52

Level **A**

1. Calculate mentally and then write the answers.

4.2 + 3.3 =   9.3 + 2.5 =   2.3 + 13.67 =

2.34 + 3.63 =   0.24 + 0.31 =   0.46 + 0.54 =

0.78 + 2.2 =   6.4 + 1.07 =   0.55 + 5.5 =

2. Use the column method to add these decimal numbers, then check the answers with a calculator. Write your answers.

2.26 + 0.9   3.25 + 7.49   0.64 + 1.48

3.6 + 15.4   1.608 + 2.84   30.84 + 57

3. Use the method shown in the shaded example to add these decimal numbers. Write the missing numbers in the brackets.

2 tonnes, 50 kilograms + 270 kilograms = (2.32 tonnes)

(2.05 tonnes)   (0.27 tonnes)

£5, 67 pence + £3, 90 pence = (   )

(   ) (   )

4 metres, 25 centimetres + 3 metres, 80 centimetres = (   )

(   )   (   )

600 ml + 9 l, 400 ml = (   )

(   ) (   )

4. Calculate the missing terms and write the answers in the table. How much are Dylan's parents' phone bills and internet bills for the last two months? (units: £)

	Phone bills	Internet bills	Sum
April	78.60	130.00	
May	80.20	130.00	
Total			

2. Recognise decimal numbers and their addition and subtraction

Level B

1. Think about these column additions and write the missing number in the boxes.

```
    6 . ☐              3 ☐ . 9
+  ☐☐ . 5          +  ☐ 7 . ☐
  ─────             ─────────
   2 0 . 1           ☐☐ 1 . 2
```

2. Practice

The Brown family had saved up £400 to spend on their holiday.
The table below shows how much they spent each day.
How much did they spend during the whole week?
How much was left over?

	Spent (£)
Monday	£67.30
Tuesday	£43.80
Wednesday	£71.00
Thursday	£19.54
Friday	£152.23
Saturday	£27.99
Sunday	£15.15

Total spent: _____

Amount left over: _____

Pupil Textbook page 53   Level A

1. Mental arithmetic. Subtract these decimal numbers.

   7.8 − 3.1 =   5.81 − 1.8 =   1 − 0.9 =
   0.83 − 0.53 =   4.68 − 2.37 =   2.94 − 2 =
   7.25 − 7.19 =   3.12 − 0.2 =   2 − 0.99 =

2. Use the column method to subtract these decimal numbers. Check the answers with a calculator.

   13.58 − 8.68 =   1.78 − 0.9 =   6.07 − 2.869 =

   7.03 − 4.57 =   1.5 − 0.306 =   90 − 9.09 =

3. Use decimal numbers to work out the answers to the following subtraction questions.

   5 kilograms − 2 kilograms, 23 grams   6 m² 400 cm² − 5400 cm²

   4 km 50 m − 244 m   £10 − 97 pence

2. Recognise decimal numbers and their addition and subtraction

4. Read each question carefully and work out the answer.

a. The school spent £160.78 on footballs and basketballs. £83.20 was spent on buying footballs. How much money was spent on buying basketballs?

b. A younger brother is 1.33 metres tall. He is 5 centimetres shorter than his older sister. How tall is his older sister?

Level B

Think about these column subtractions and write the missing numbers in the boxes.

```
    4 . ☐ ☐              8 ☐ . 2
  - ☐ . 9 0 7          -   9 . ☐ 1
  ─────────            ─────────
    0 . 2 4 3            7 ☐ . 4 9
```

Pupil Textbook pages 54–55

Level **A**

1. Calculate mentally and then write the answers.

3.5 + 2.1 =   1 − 0.2 =

0.58 + 0.32 =   7 + 3.4 =

0.9 − 0.1 =   8.26 + 0.4 =

12.1 − 7.6 =   0.6 − 0.06 =

2. These column calculations contain mistakes. Think about where the mistakes are, and then correct the calculations.

```
      7 . 8 8              6 . 7              5 . 0 6
+ 1 3 . 2 2            − 2 . 8 3          −   4 . 2
─────────              ─────────          ─────────
    2 0 . 1 0              4 . 9 7                8 6
```

The correct calculation: The correct calculation: The correct calculation:

_____   _____   _____

3. Use the column method to work out these additions and subtractions, then check your answers with a calculator. Write the answers in the spaces below.

76.8 + 26.82   30.49 − 15.9

10 − 3.85   25.903 + 4.097

2 Recognise decimal numbers and their addition and subtraction

4. Write these fractions as decimal numbers. Then work out the answers.

$\frac{2}{10} + \frac{3}{10}$

$\frac{9}{10} + \frac{4}{100}$

$\frac{80}{100} - \frac{56}{100}$

$\frac{7}{10} - \frac{18}{100}$

5. Use decimal numbers to calculate the answers to these addition and subtraction questions.

6 km − 3 km 580 m

£4, 62 pence + £3, 8 pence

4 tonnes 60 kg − 870 g

6 m² − 2 m² 6300 cm²

6. Multiple choice – write the letter of the correct answer in the brackets.

a. Dylan's father took £500 from his bank account to buy a new laptop. It cost £365.20. How much was left? The correct number sentence is (   ).

   A. 500 − 365.2 = 34.8   B. 500 − 365.2 = 145.8
   C. 500 − 365.2 = 134.8

b. Poppy bought a book for £12.30 and then bought an exercise book for £1.80. She spent (   ) in total.

   A. £14.10   B. £14, 1 pence   C. £14.01

7. **Read each question carefully and work out the answer.**

   a. A café bought 2.1 tonnes of rice in December, which was 0.3 tonnes more than in November. How many tonnes of rice were purchased in November?

   b. Some pupils collected leaves. The pupils of Year 4, Class 1 collected 32.56 kilograms and the pupils of Year 4, Class 2 collected 39.54 kilograms. How many kilograms have the pupils in the two classes collected?

Level B

1. The greatest pure decimal number that consists of 0, 4, 5 and 9 is (   ) and the smallest is (   ). The sum of the two numbers is (   ). The difference between the two numbers is (   ).

2. A rope is folded in half, folded in half again and then folded in half for a third time. The length of each segment is 0.5 metres. The length of the whole rope is (   ) metres.

2 Recognise decimal numbers and their addition and subtraction

8. The application of addition and subtraction of decimal numbers

Pupil Textbook pages 56–59 *Level A*

1. Write a number in each ☐ to make the calculations easier.

a. (14.15 + 5.87) + 5.85 = 14.15 + ☐ + 5.87

b. 23.02 + 13.76 + 0.98 + 6.24 = (☐ + ☐) + (☐ + ☐)

c. 28.93 − 7.46 − 5.54 = 28.93 − (☐ + ☐)

d. 8.07 − 5.8 + 0.93 = 8.07 + ☐ − 5.8

2. Calculate mentally and then write the answers.

0.48 + 5.5 =	6.2 − 2.8 =	4 − 1.4 − 1.6 =
10.74 − 0.4 =	0.04 + 0.4 =	3.2 + 9.4 + 6.8 =
4.3 − 1.6 =	1.69 − 0.69 =	14.3 − 5.6 − 4.3 =

3. Compare the answers to these calculations and write <, = or > in each ◯.

10 − 2.7 − 7.3 ◯ 10 − (7.3 − 2.7)

87.5 − 63.9 + 25 ◯ 87.5 + 25 − 63.9

52.8 − 38.6 + 1.2 ◯ 52.8 − (38.6 + 1.2)

63.4 + 27.5 − 7.5 ◯ 63.4 + (27.5 − 7.5)

4. Work these out, showing the steps in your calculation. Use a shortcut strategy to calculate quickly.

5.8 + 6.39 + 3.61   13.56 − 1.48 − 3.52

10.75 + (0.4 + 0.25)   4.587 + 3.47 − 2.47

7.4 + (5.96 − 3.86)   13.7 + 1.94 + 0.06 + 5.3

20.89 − (4.56 + 13.89)   38.14 − 5.86 − 4.14

38.4 + (24.24 − 18.4)   79.8 − 5.8 + 4.2

5. Calculate the perimeter of each shape. (units: cm)

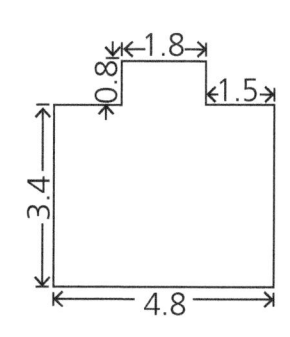

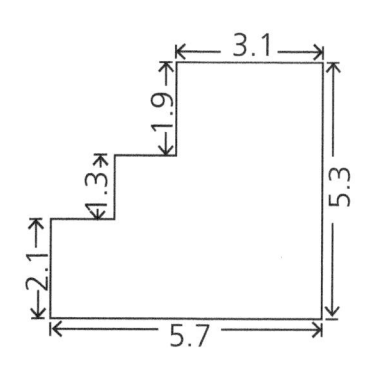

2. Recognise decimal numbers and their addition and subtraction

Level B

1. Dylan is planning a small party. He wants to buy garlic bread, pizza, chicken wings and a drink (one of each item). The price list shows how much the items would cost at two takeaways near Dylan's home. He wants to buy all the things he wants at the same takeaway. Help him to decide which takeaway to use.

Food item	Price at takeaway A (£)	Price at takeaway B (£)
garlic bread	6.90	6.70
pizza	23.95	21.25
chicken wings	13.20	12.60
drinks	5.10	5.40

   a. I suggest that Dylan should use takeaway _____.

   b. My reason is that _____
   _____.

   c. Dylan takes £50. He should have £ _____ left.

2. Work out the following additions and subtractions and write the answers. Use a shortcut strategy to calculate quickly.

   9.9 + 99.9 + 999.9 + 9999.9   10 − 0.1 − 0.3 − 0.5 − 0.7 − 0.9

   $\frac{23}{100}$ + 100.01 + 0.77 − $\frac{1}{100}$

9. Practice exercise (2)

Pupil Textbook page 60 Level **A**

1. Work these out, showing the steps in your calculation. Use a shortcut strategy to calculate quickly.

   19.48 − 2.36 + 8.52 − 4.64   32.86 − 7.64 + 10.36

   2.48 + (5.89 + 3.52) + 4.11   40.36 − (26.4 + 10.36)

   47.35 + 5.74 + 16.65 + 18.26   38.46 + 2.77 − 38.46 + 2.77

2. Read each question carefully and work out the answer.

   a. These are the populations of five continents in 2009.
      Europe: 0.732 billion   Asia: 4.12 billion   America: 0.931 billion
      Africa: 1.01 billion   Oceania: 0.03 billion

      i. What is the difference between the highest population and the lowest population in 2009?

      ii. What was the total population of the five continents in 2009?

      iii. What other information can you learn from the calculations?

2 Recognise decimal numbers and their addition and subtraction

b. The number of visitors to a scenic spot was 1.56 thousand on Thursday and 1.68 thousand on Friday. Compared to the total number of people on the previous two days, there were 0.56 thousand fewer on Saturday. How many people were there on Saturday?

c. Pupils prepare streamers for a party. The red streamer is 58.7 metres long, which is 4.3 metres longer than the yellow streamer. The green streamer is 5.7 metres shorter than the yellow streamer. How long is the green streamer?

Level B

An engineer has two wires. The first one is 48.3 metres long, which is 6.4 metres longer than the second. After the engineer has used 9.6 metres of the first one, how many metres shorter is it than the second wire?

Unit Three: Statistics

> Record the temperature outside your school every day at the same time for a week and draw a line graph.

The table below lists the sections in this unit.

After completing each section, assess your work.

(Use 😊 if you are satisfied with your progress or 😐 if you are not satisfied.)

Section	Self-assessment
1. Introduction to line graphs	
2. Draw a line graph	

3 Statistics

1. Introduction to line graphs

Pupil Textbook pages 62–68

Level A

1. Read the line graph and answer the questions.

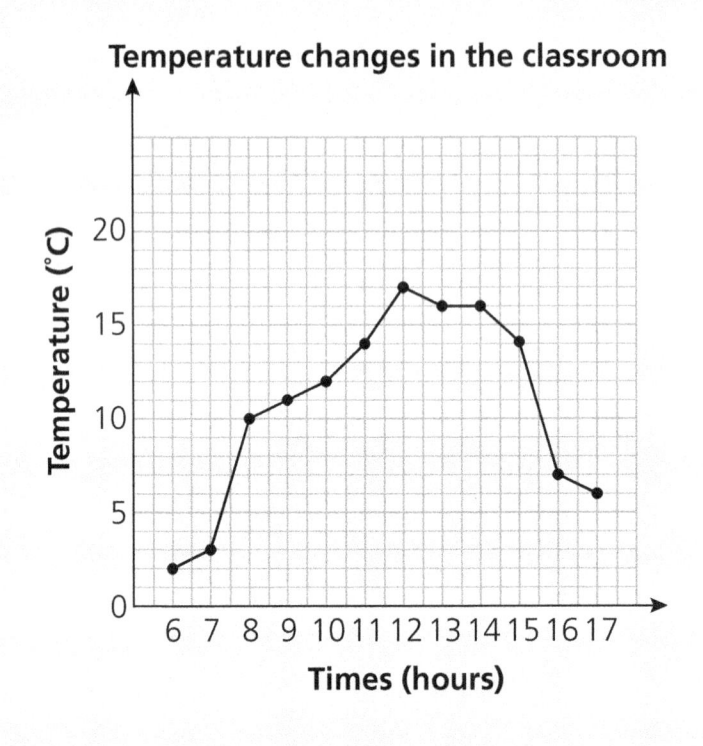

Temperature changes in the classroom

a. The line graph data was measured every _____ hours.

b. The classroom temperatures at 7:00 and 14:00 were _____ °C and _____ °C.

c. The classroom temperature was the highest at _____, and it was _____ °C.

d. The maximum increase in classroom temperature was between _____ and _____.

e. There was no change in temperature between _____ and _____.

f. There were _____ hours when the temperature was above 10 °C in the classroom.

2. Look at these pairs of line graphs. Which of the two is better in each pair? Mark with a '✓' and explain your reasoning.

a. Dylan's body temperature

Dylan's body temperature

Graph A (   )   Graph B (   )

Reason: _____

b. Green garden area per person in Shanghai

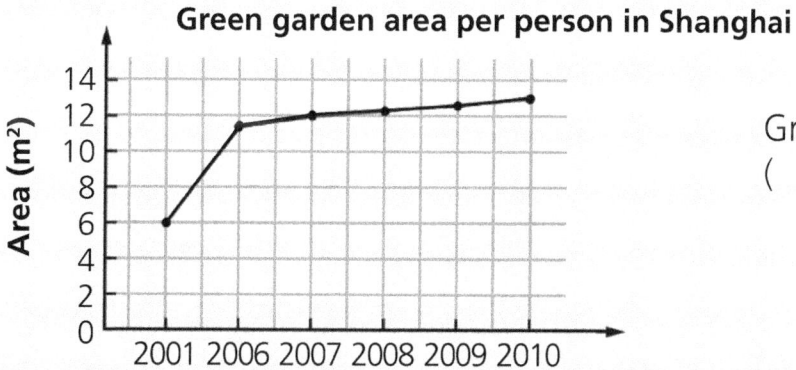

Graph A (   )

Green garden area per person in Shanghai

Graph B (   )

Reason: _____

3 Statistics

3. Read the line graph and answer the questions.

Changes in Emma's mass from last April to April this year

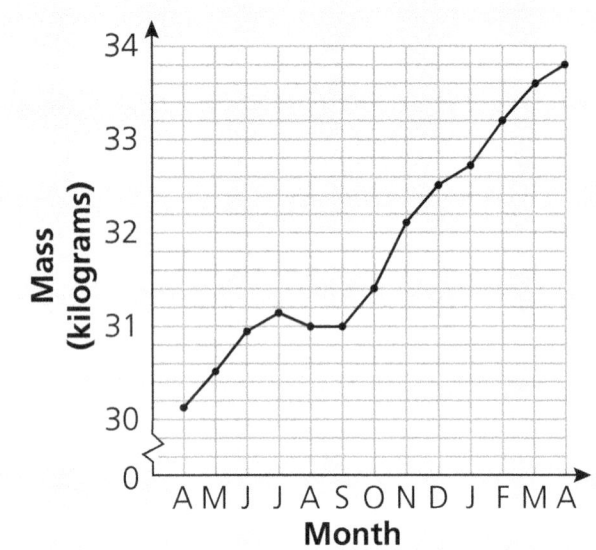

a. i. In the line graph, what does the horizontal axis represent?

   ii. What does the vertical axis represent?

b. What does a single unit on the vertical axis stand for (in kilograms)?

c. What was Emma's mass last July?

d. When did Emma reach 33 kilograms?

e. When was Emma's mass less than it was the month before?

f. From which month to which month did Emma's mass stay the same?

g. From which month to which month did Emma's mass show the greatest increase?

4. Answer the questions based on the line graph.

A patient's body temperature from 6:00 on 8 June to 18:00 on 10 June

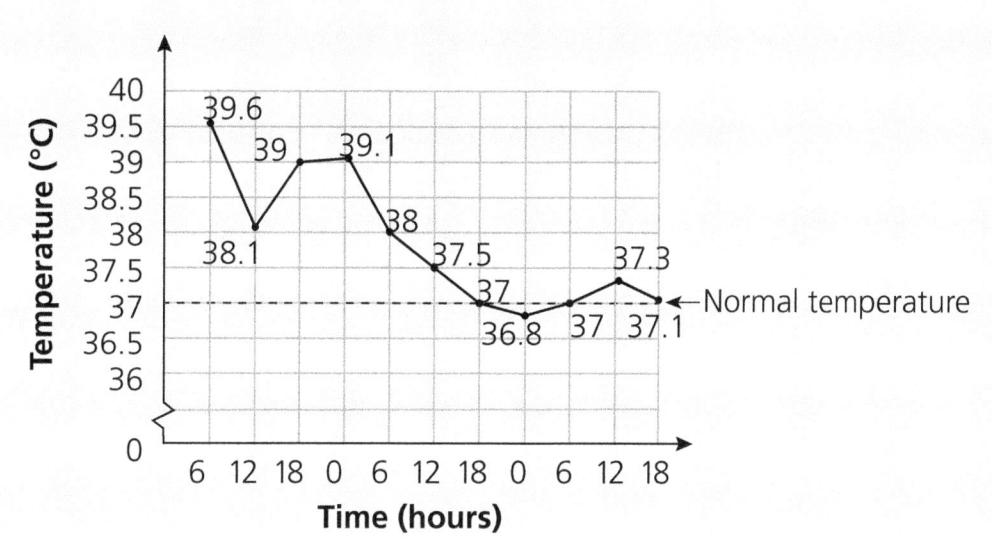

a. The horizontal axis represents _____; the vertical axis represents _____.

b. The nurse checks the patient's temperature every _____ hours.

c. The patient's temperature at 18:00 on 9 June was _____ °C.

d. At what time did the patient have the highest body temperature?

e. When did the patient's temperature go down the fastest?

f. What does the zigzag line mean? What does the dotted line mean?

g. Did the patient get better or worse?

3 Statistics

Level B

Look at the line graph and answer the following questions.

The graph on the right shows the number of people visiting a supermarket between 17:00 and 22:00.

Number of people from 17:00 to 22:00 in a supermarket

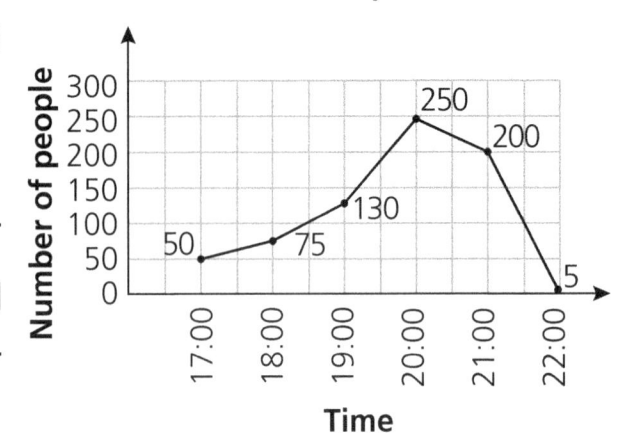

The number of people increased the most from _____ to _____.

The number of people decreased the most from _____ to _____.

Based on the number of people in the supermarket, what are your suggestions for organising the number of cashiers in the supermarket?

My suggestions:

_____
_____
_____

2. Drawing a line graph

Pupil Textbook pages 69–72　　　　Level A

1. Draw a line graph.

a. The table shows the numbers of leaves growing during the 13 days after the budding of a plant.

Days	1	2	3	4	5	6	7	8	9	10	11	12	13
Number of leaves	2	2	3	3	4	4	5	6	9	11	13	17	20

Draw a line graph based on the data in the table.

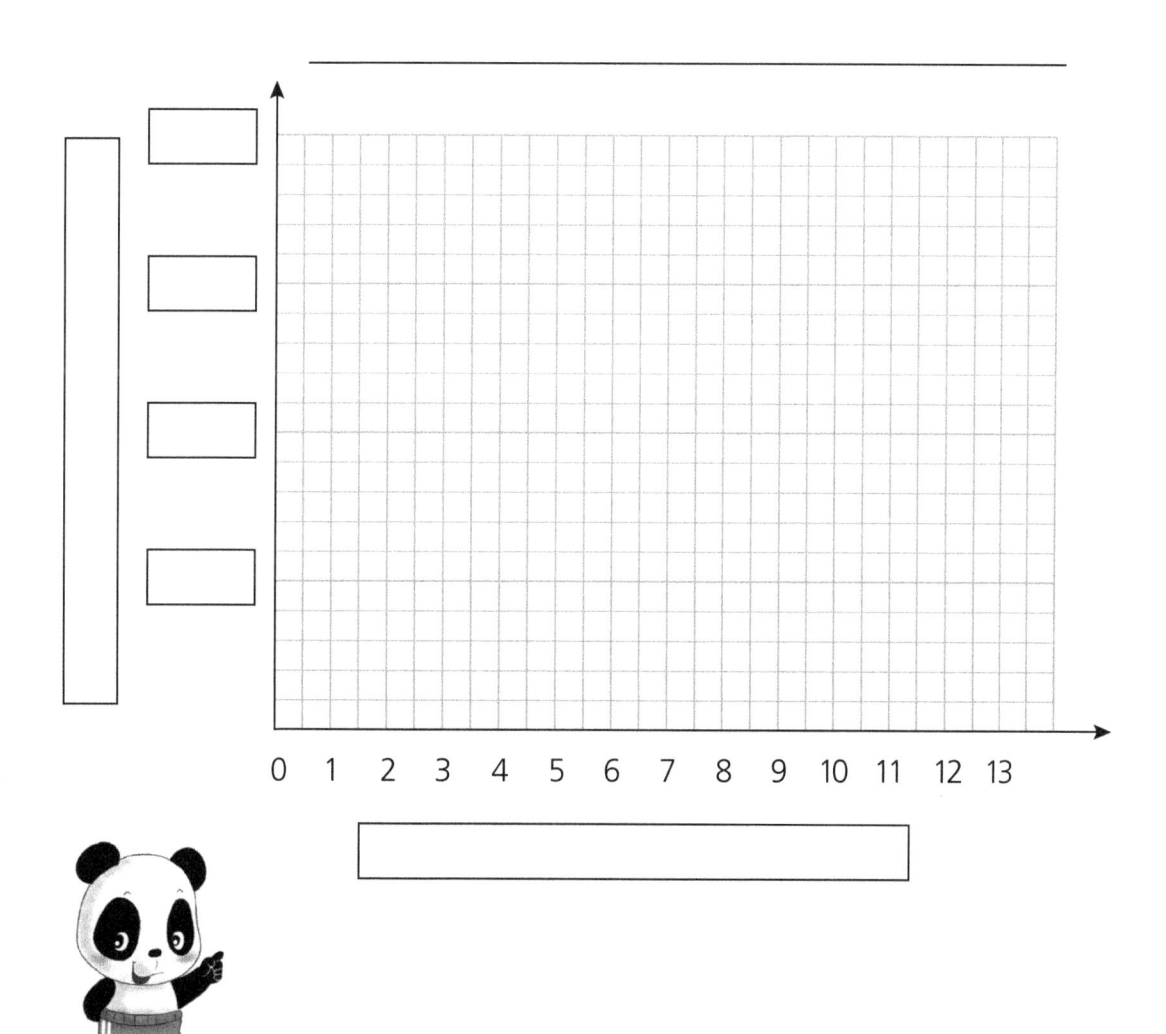

3 Statistics

b. Alex measured the temperature of the swimming pool water every 2 hours, and recorded his data in the table below. Use the information in the table to draw a line graph to show how the temperature of the water in the swimming pool changed.

Temperature change of water in the swimming pool

Time (hours)	Temperature (°C)
6	22.5
8	24
10	26.5
12	27.5
14	29
16	26.5
18	25.5
20	24

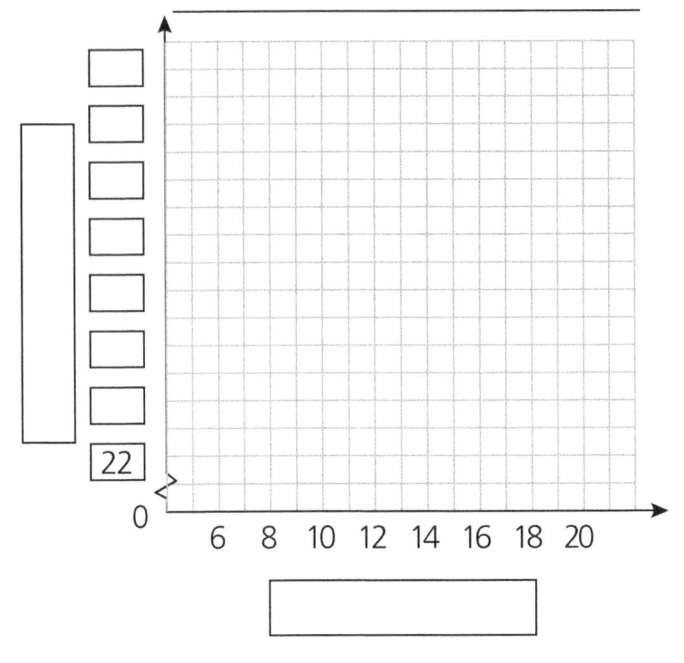

c. This table shows changes in Sunil's weight. Use the information in the table to draw a line graph.

Sunil's weight

Month	Kilograms (kg)
Apr	23.8
May	24.5
Jun	24.7
Jul	24.0
Aug	24.3
Sep	25.1

(measured on the 15th of each month in 2015)

Level B

1. Class 4B use the internet to investigate the number of children born in their city from 2007 to 2012, and find the following data. Use the information in the table to draw a line graph.

Investigation into the number of children born in our city from 2007 to 2012

Year	Number of births (thousands)
2007	10.1
2008	9.7
2009	9.2
2010	10.0
2011	10.2
2012	12.1

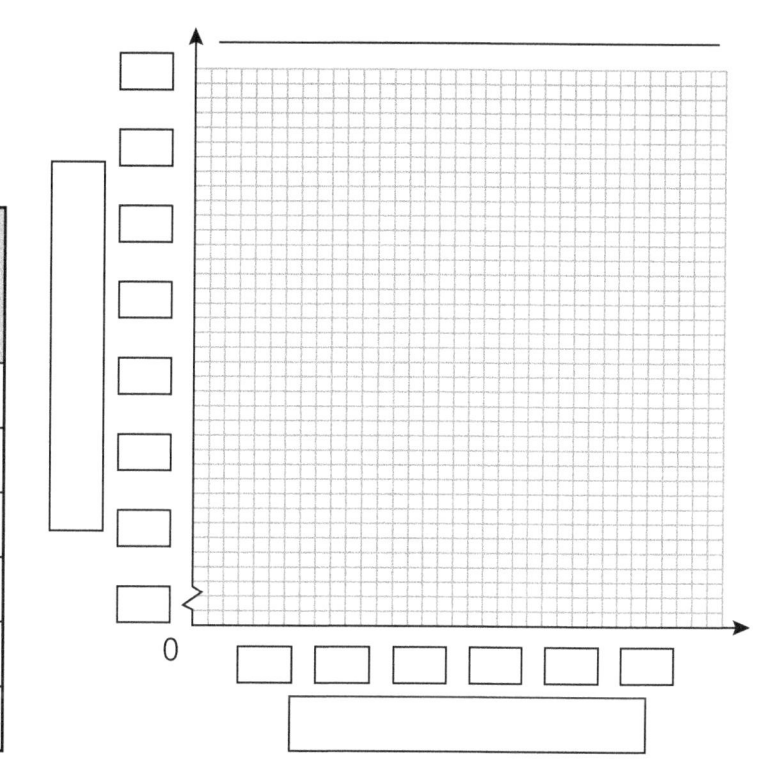

Talk about your ideas after reading the line graph.

_____
_____
_____

3 Statistics

2. Record the temperature outside your school every day at the same time for a week and draw a line graph based on the data.

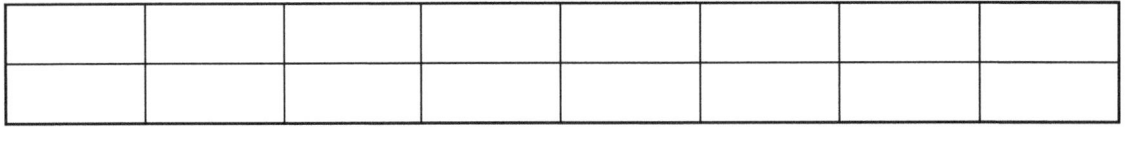

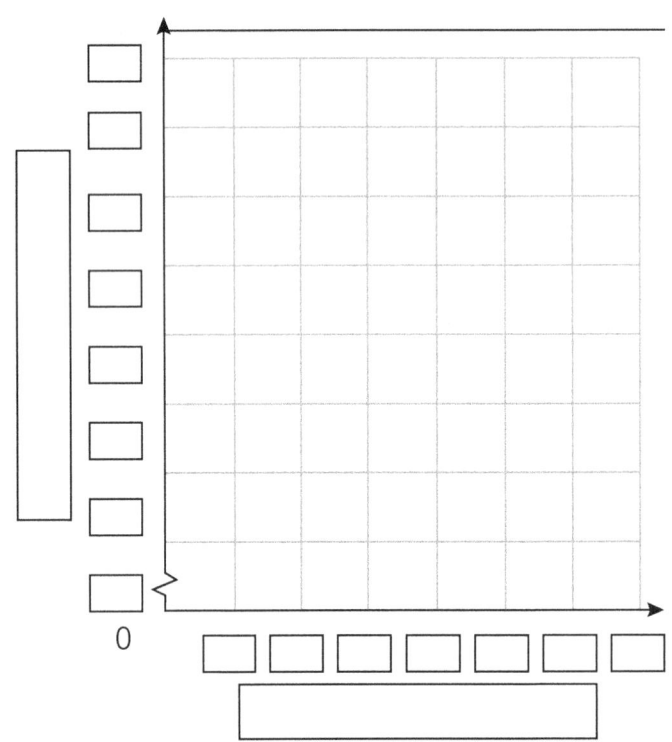

Use your understanding of temperature and the information in the line graph to talk about what you found out.

_____

_____

Unit Four: Measure and geometry

Uncover the secret of buried treasure. (The clue is in this lesson.)

The table below lists the sections in this unit.

After completing each section, assess your work.

(Use 🙂 if you are satisfied with your progress or 😐 if you are not satisfied.)

Section	Self-assessment
1. Perpendicular	
2. Parallel	
3. Practice exercise (3)	

4 Measure and geometry

1. Perpendicular

Pupil Textbook pages 74–75   Level A

1. Fill in the missing words.

   a. When two straight lines intersect at right angles, these two lines are _____ to each other, and one of the lines is called a _____ line to the other line. The point where the two lines _____ is called the point of intersection.

   b. The top and side of a rectangle are _____ to each other.

2. Multiple choice – write the letter of the correct answer in the brackets.

   a. Which diagram shows a pair of lines that are perpendicular to each other? (   )

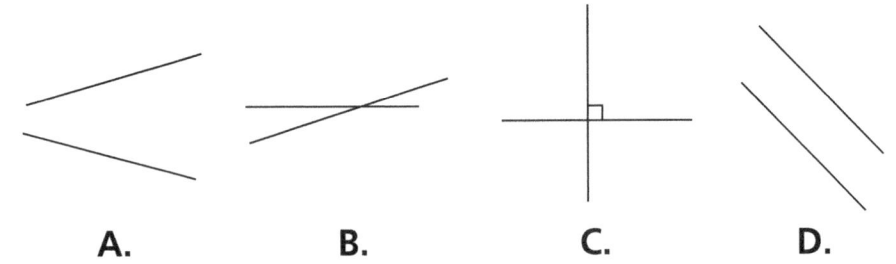

   A.   B.   C.   D.

   b. At (   ), the hour hand and the minute hand on a clock face are perpendicular to each other.

   A. 6:00   B. 3:30   C. 9:00   D. 12:15

3. True or false? Put a tick (✓) for 'true' or a cross (✗) for 'false' in the brackets.

  a. When two straight lines intersect at right angles, these two lines are perpendicular to each other. ( )
  b. The point where the two lines intersect is called the point of intersection. ( )
  c. Two perpendicular lines form only one right angle. ( )
  d. There are three pairs of perpendicular sides in a set square. ( )

Level B

There are ( ) pairs of perpendicular lines in this diagram.

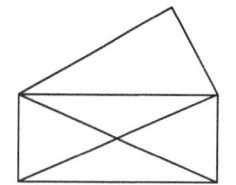

4 Measure and geometry

Pupil Textbook pages 76–79 Level A

1. **Multiple choice – write the letter of the correct answer in the brackets.**

   a. You can draw ( ) line(s) perpendicular to a line from a point outside it.
      A. 1   B. 2   C. 3   D. infinitely many

   b. If you draw a perpendicular line b to another line a from a point outside line a, the greatest possible number of right angles formed will be ( ).
      A. 1   B. 2   C. 3   D. 4

   c. In the diagram, the shortest line segment from point A to line segment BE is ( ).
      A. AB B. AC
      C. AD D. AE

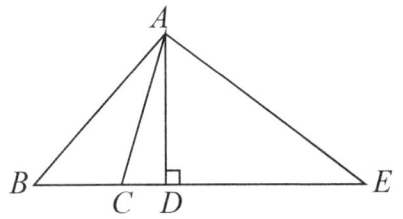

2. **True or false? Put a tick (✓) for 'true' or a cross (✗) for 'false' in the brackets.**

   a. There are an infinite number of lines perpendicular to one given line.   (　)

   b. Of all the line segments from a point outside a line to the line itself, the shortest line segment is the perpendicular from the point to the line.   (　)

   c. When two lines intersect at right angles, they are perpendicular.   (　)

   d. At 9:30, the hour hand and the minute hand on a clock face are perpendicular to each other.   (　)

   e. In a rectangle, only two pairs of adjacent sides are perpendicular.   (　)

   f. If line $a \perp b$, line a and line b are perpendicular.   (　)

3. Measure the distance from point A to line l. Draw your answers on the diagrams.

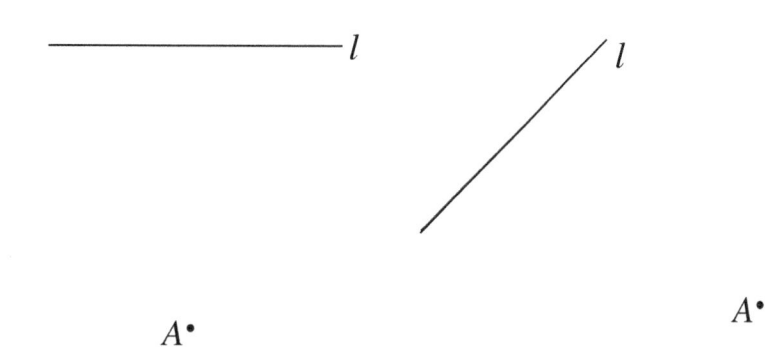

4. In this diagram the distances from points A, B and C to line l are (　　), (　　) and (　　), respectively.

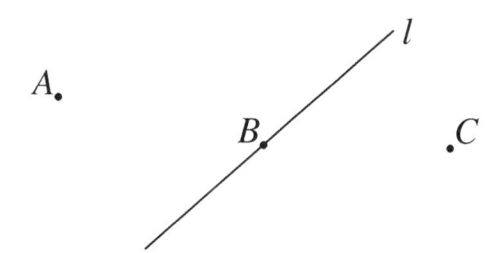

5. Draw two pairs of perpendicular lines on the grid: $a \perp b$ and $c \perp d$.

4 Measure and geometry

6. In the diagram, ∠1 = 40°, ∠2 = 50°. Are line *a* and line *b* perpendicular? Explain your reasoning.

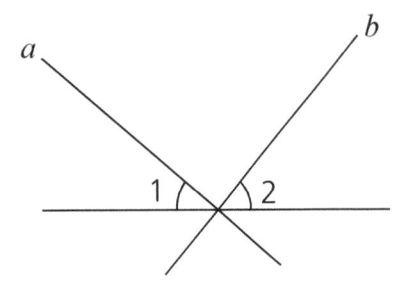

7. Draw a rectangle 3 cm long and 2 cm wide.

Level B

1. When measuring the long jump, which of the line segments in the diagram on the right would give the correct length? Why?

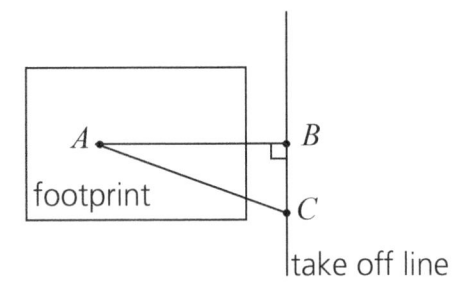

2. Alex plans to go fishing on the river bank. He wants to walk the shortest distance. Which route should he take? Draw the route on the diagram below.

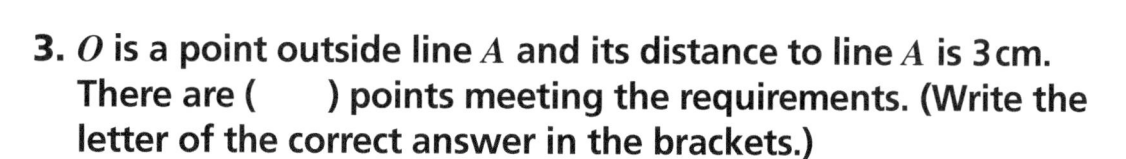

3. O is a point outside line A and its distance to line A is 3 cm. There are (   ) points meeting the requirements. (Write the letter of the correct answer in the brackets.)

    A. 1   **B.** 2   **C.** infinitely many

4 Measure and geometry

2. Parallel

Pupil Textbook page 80 Level A

1. Look at each diagram and decide which lines are parallel. Write the answers in the brackets.

a.

( ) ∥ ( )

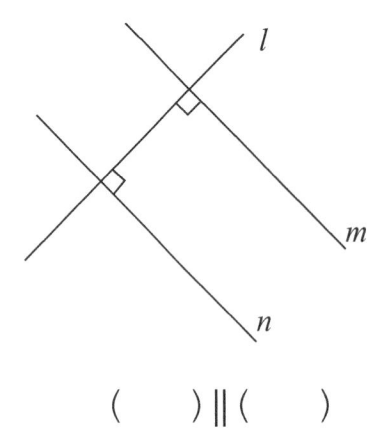

b.

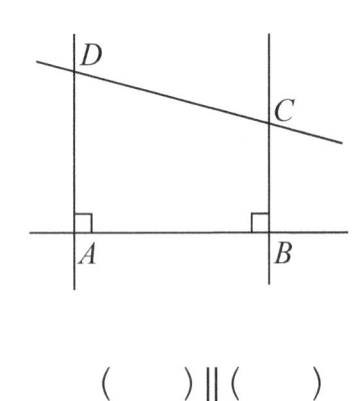

( ) ∥ ( )

c.

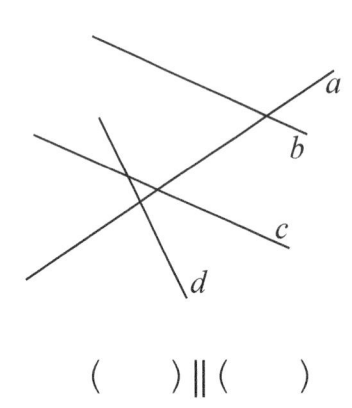

( ) ∥ ( )

d.

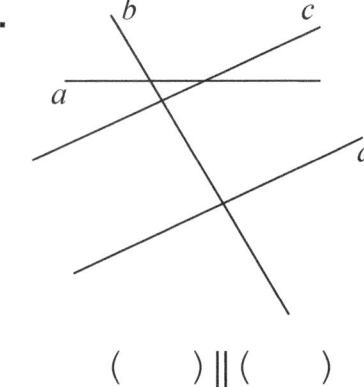

( ) ∥ ( )

2. Read the problems and write the answers in the brackets.

a. On the same piece of paper, if lines a and b are perpendicular to the same line, then line a and line b are (　　) to each other and this can be denoted as (　　).

b. There are (　) pairs of parallel sides in a rectangle.

3. Multiple choice – write the letter of the correct answer in the brackets.

   a. There are ( ) pairs of parallel lines in the diagram on the right.
      A. 1   B. 2
      C. 3   D. 4

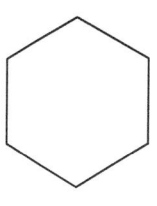

   b. In the square on the right there are ( ).
      A. 2 pairs of perpendicular lines, 4 pairs of parallel lines 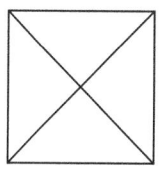
      B. 4 pairs of perpendicular lines, 2 pairs of parallel lines
      C. 5 pairs of perpendicular lines, 2 pairs of parallel lines
      D. 6 pairs of perpendicular lines, 2 pairs of parallel lines

4. How sharp are your eyes? Are lines a and b in these two diagrams parallel? Use tools to check.

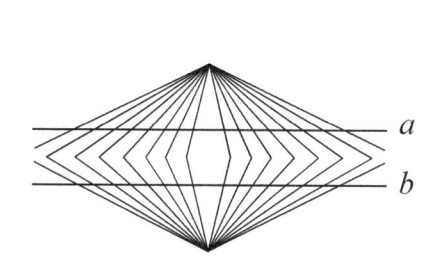

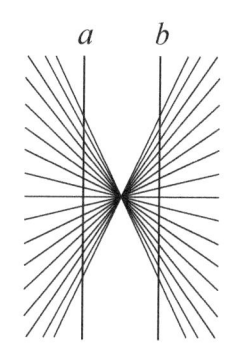

Level B

1. a, b and c are three lines on the same sheet of paper. If $a \perp b$ and $a \parallel c$, then b ( ) c.

2. There are ( ) sets of parallel lines in this diagram.

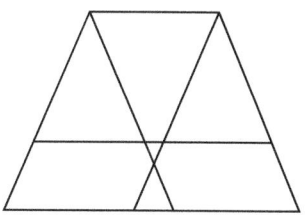

4 Measure and geometry

Pupil Textbook pages 81–84 Level A

1. In each diagram, draw the line that is parallel to the line l through point A.

2. Draw two more lines b and d on the grid such that: $a \parallel b$, $c \parallel d$.

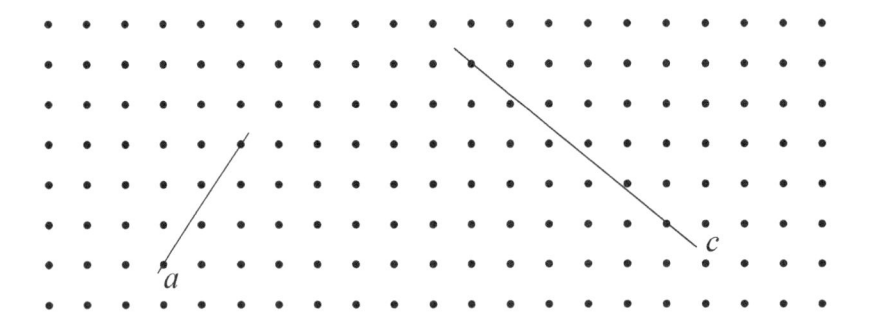

3. Measure the distances between AB and DC, AD and BC in the rectangle and mark them on the diagram.

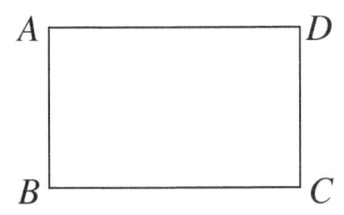

4. In this diagram, line a is parallel to line b. Measure the distance between lines a and b. Draw and mark it on the diagram.

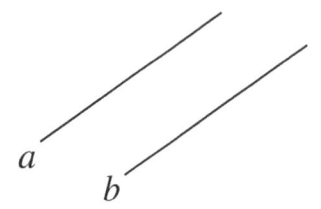

5. Multiple choice – write the letter of the correct answer in the brackets.

   a. There are (   ) of parallel lines in the diagram on the right.
      A. 2 pairs   B. 3 pairs
      C. 4 pairs   D. 5 pairs

   b. (   ) lines can be drawn parallel to a straight line.
      A. 1   B. 2   C. 3   D. infinitely many

6. The diagram shows two parallel water pipes. They are to be connected by the shortest possible length of piping. Which line segment, a, b, c or d, shows the best place to lay the new pipe? Explain your reasoning.

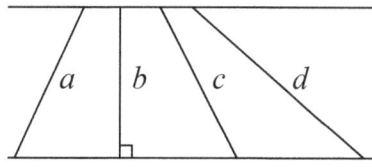

7. In each of these diagrams, draw lines through point P that are parallel to the two sides of the angle.

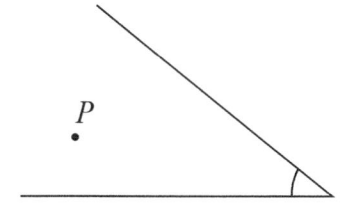

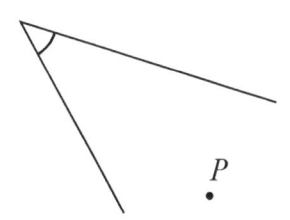

8. On this diagram, draw a line parallel to the line segment OB from point A and a line parallel to line segment AO from point B.

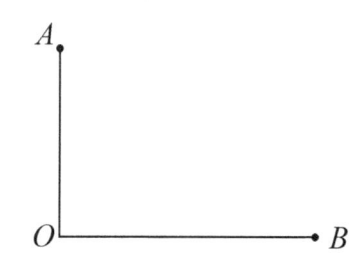

4 Measure and geometry

Level B

1. Draw lines parallel to line a with a distance between them of 2 cm. How many lines can you draw?

2. Draw a line parallel to line segment OB from point M. Draw a line parallel to line segment OA from point N. The two lines intersect at point P. Measure the distance from point P to OA and OB and write the measurements on the diagram.

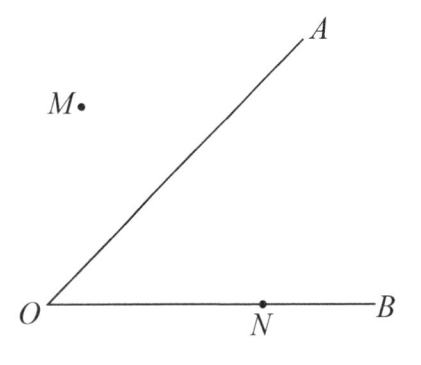

3. Practice exercise (3)

Pupil Textbook pages 85–86 Level A

1. The number shapes below are made of lolly sticks. How many pairs of parallel lines are there in each diagram?

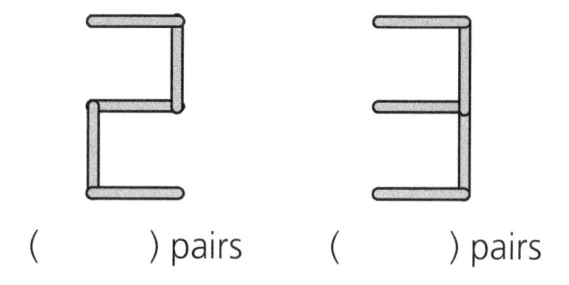

   (   ) pairs   (   ) pairs

2. Change this diagram into a rectangle by drawing parallel lines.

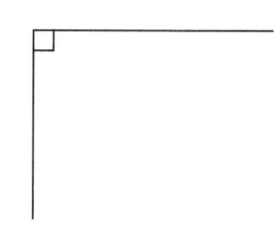

3. Draw two parallel lines and make the distance between them 20 mm.

4. Draw and measure these lines and distances.
   a. Draw a line perpendicular to line l from point P.
   b. Draw a line parallel to line l through point P.
   c. Measure the distance from point P to line l. The distance between the two parallel lines is _____.

   • P

   l ─────────────

4 Measure and geometry

Level B

1. The lines a, b and c are drawn on a sheet of paper, with $a \parallel b$ and $b \parallel c$. The distance between lines a and b is 2 cm and the distance between lines b and c is 5 cm. What could the distance between lines a and c be?

2. Make a drawing from the information you are given.

An explorer found a treasure map, but the only things marked on it are three points, A, B and C, and a few words.

> There is a straight river passing through points B and C. The treasure is located above and to the right of point A. The distance between the treasure and point A is the same as the distance from point A to the river. The line between point A and the location of the treasure is parallel to the river.
>
> A •
>
> • C
>
> B •

Can you find the location of the treasure accurately?

Unit Five: Consolidating and enhancing

In one year 5 hives of bees produced 350 kilograms of honey. The next year, the amount of honey increased by one time.

How many kilograms of honey will there be?

The table below lists the sections in this unit.

After completing each section, assess your work.

(Use 😊 if you are satisfied with your progress or 😐 if you are not satisfied.)

Section	Self-assessment
1. Problem solving (2)	
2. Decimal numbers and approximate numbers	
3. Perpendicular and parallel	

5 Consolidating and enhancing

1. Problem solving (2)

Pupil Textbook pages 88–89 Level A

1. Look at the pictures. Work out the number of apples and write the answers in the brackets.

   a. original number: 🍎🍎

   number now: 🍎🍎 🍎🍎 🍎🍎

   Think of the original apples as one part. Now there are (   ) parts. The number of parts has increased (   ) times compared to the original. It can also be said that the number of apples has increased to (   ) times the original. The equation to work out 'how many apples there are now' is (   ).

   b. original number: 10 🍎

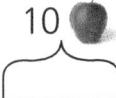

   number now: ? 🍎

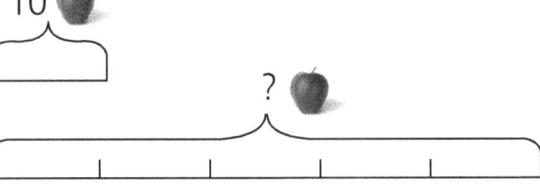

   Think of the original apples as one part. Now the number of apples has increased to be (   ) times the original number. It has increased (   ) times. The equation to work out 'how many apples there are now' is (   ).

2. Work out the answer and draw it.

   a. Start with △ △ △ △
   Increase by 2 times: _____
   Increase to 2 times: _____
   Increase by 2 times and then by 1 time:
   _____

b. Start with ○○○

Increase by 3 times: _____

Increase to 4 times: _____

Increase by 4 times but subtract 1 time:

_____

3. Fill in the brackets.

a. There are 20 🌼.

If the number increases by 2 times, there will be (   ) 🌼.

If the number increases to 2 times, there will be (   ) 🌼.

b. There are 10 🥚. If the number increases by (   ) times there will be 50. If the number increases to (   ) there will be 50. Increasing (   ) gives 80.

4. Use the column method of calculation to work out the answers.

a. a is 52. If a increases to 3 times, it will equal b. What is b?

b. a is 52. If a increases by 5 times and then 2 times is subtracted, it will equal b. What is b?

Level B

Dylan has 18 🍎.   Emma has 9 🍎.

Alex has 6 🍎.   Poppy has 36 🍎.

If (   )'s 🍎 increase by (   ) times, there will be as many as (   )'s 🍎.

(   )'s 🍎 increase to (   ) times, there will be as many as (   )'s 🍎.

5 Consolidating and enhancing

Pupil Textbook pages 90–93　　　Level A

1. Comparison practice

a. i. One year, 5 hives of bees produced 350 kilograms of honey. The next year, the amount of honey increased by one time. How many kilograms of honey will this be?

ii. 5 hives of bees can produce 350 kilograms honey per year. The number of hives is increased by 5 times. How many kilograms of honey will there be?

b. i. The school feeds 8 rabbits. The rabbits eat 8 kilograms of carrots in total per day. The number of rabbits is increased by 2 times. How many kilograms of carrots will they eat per day?

ii. The school feeds 8 rabbits. The rabbits eat 8 kilograms of carrots in total per day. The number of rabbits is increased to 2 times the number. How many kilograms of carrots will they eat per day?

2. Multiple application

a. Read the conversation between Dylan, Dylan's father and Dylan's grandfather. How old is Dylan? How old is Dylan's father?

> My age increased by 3 times is my father's age.

> My age increased to 2 times is my father's age.

> I'm 80 years old!

b. A new road is being built. 240 kilometres have already been built. When it is finished, the new road will be 8 kilometres less than 4 times the total length already built. How many kilometres of road are still to be built?

c. The total world population was about 2.5 billion in 1950. It is estimated that the population in 2050 will be be 2 billion more than 3 times the population in 1950. What will the population be in 2050?

5 Consolidating and enhancing

d. A gardener is going to mow a lawn with an area of 2400 square metres. It is estimated that 75 square metres will take 1 hour. The gardener's actual work rate is increased by 2 times. How long will it actually take? How much less than the estimated time is this?

e. Recycling 3 tonnes of waste paper means that 50 trees don't need to be cut down. 12 tonnes of waste paper is recycled. How many trees will be saved from being cut down?

Level B

Some road builders planned to repair 1944 metres of road in 12 days. They were asked to reduce this to half that time. How many extra metres per day do they need to repair now?

2. Decimal numbers and approximate numbers

Pupil Textbook pages 94–99　　　Level A

1. Follow the instructions in the table to round the numbers.

	Round the number to the nearest whole number	Round the number to the nearest tenth	Round the number to the nearest hundredth
70.047			
0.599			
5.783			

2. Follow the instructions in the table to round the numbers to the nearest hundredth.

	Round down	Round up	Round (4 or less rounds down, 5 or more rounds up)
5.495			
0.996			
18.1843			

5 Consolidating and enhancing

3. First, write each number in billions. Then round your answers to the nearest hundredth.

These are the passenger capacities of different modes of transport in China for the year 2012.

Rail:
1 893 368 500 passengers
_____
_____

Bus:
35 570 100 000 passengers
_____
_____

Ship:
257 520 000 passengers
_____
_____

Aeroplane:
319 360 500 passengers
_____
_____

4. Round the numbers as instructed and write the answers in the brackets.

a. Zhao Zhou Bridge is a masterpiece among Chinese ancient architectural bridges. Its length is 50.82 metres. Rounding the number to the nearest 0.1, it is (   ) metres. Rounding to the nearest whole number it is (   ) metres.

b. There is a type of small horse called a Falabella. It is only 38.1 centimetres tall. Rounding the number to the nearest whole number, this is (   ) centimetres. Its weight is only 9.07 kilograms. Rounding the number to the nearest tenth, this is (   ) kilograms.

c. On 29 January 2013, one US dollar could be be exchanged for 6.2806 yuan. Then 10 US dollars could be exchanged for (   ) yuan. Ten thousand US dollars could be exchanged for (   ) yuan.

d. Rounding 19.953 to the nearest hundredth gives (   ). Rounding down to the nearest tenth gives (   ). Rounding it up to the nearest whole number gives (   ).

5. Dylan's dad had a piece of steel wire that was 14.8 metres long. He divided it into 100 equal pieces. How long was each piece? (Round the number to the nearest hundredth.)

6. Complete the table by rewriting the numbers in thousands. Then round your answers to the nearest whole number.

Date	1 Oct	2 Oct	3 Oct	4 Oct	5 Oct
Number of visitors	25 400	39 180	44 750	43 130	52 490
Number of visitors in thousands	25.4				
Rounded to the nearest whole number	25				

5 Consolidating and enhancing

7. Follow the instructions to work out the numbers. Write them in the brackets or circle them on the number line.

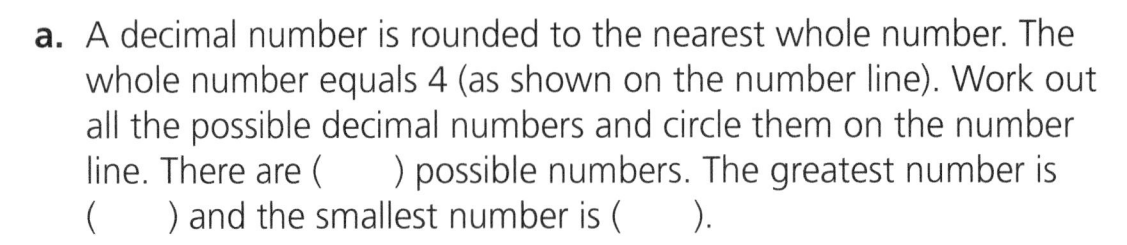

a. A decimal number is rounded to the nearest whole number. The whole number equals 4 (as shown on the number line). Work out all the possible decimal numbers and circle them on the number line. There are (   ) possible numbers. The greatest number is (   ) and the smallest number is (   ).

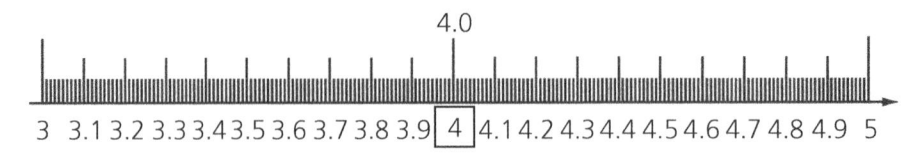

```
                         4.0
  |—————————————————————————————————————————|
  3  3.1 3.2 3.3 3.4 3.5 3.6 3.7 3.8 3.9 [4] 4.1 4.2 4.3 4.4 4.5 4.6 4.7 4.8 4.9  5
```

b. A decimal number with 2 decimal places rounded to the nearest tenth is 4.0 (as shown on the number line). The original decimal number can range from (   ) to (   ).

```
                         4.00
  |—————————————————————————————————————————|
  3  3.1 3.2 3.3 3.4 3.5 3.6 3.7 3.8 3.9 [4] 4.1 4.2 4.3 4.4 4.5 4.6 4.7 4.8 4.9  5
```

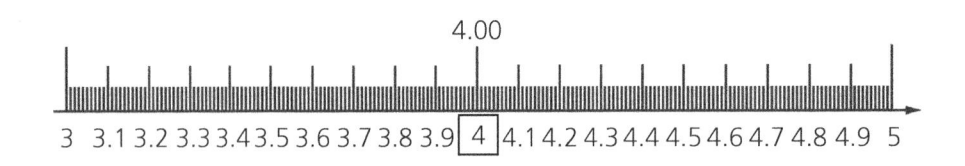

c. A decimal number with 3 decimal places rounded to the nearest hundredth is 4.00. The original decimal number can range from (   ) to (   ).

8. Multiple choice – write the letter of the correct answer in the brackets.

   a. How many numbers with 1 decimal place can be rounded up to 1? (   )
      A. 9   B. 10   C. 11   D. infinitely many

   b. How many numbers with 1 decimal place can be rounded down to 1? (   )
      A. 9   B. 10   C. 11   D. infinitely many

   c. How many numbers with 3 decimal places are more than 0.4 and less than 0.5? (   )
      A. 9   B. 99   C. 999   D. infinite

   d. Which of the following numbers is closest to 3.673? (   )
      A. 3.6735   B. 3.67309   C. 3.67297   D. 3.67315

Level B

The numbers in the top row of the table are approximate values after rounding. For each approximate number write the range of the original number. Each original number has one more decimal place than the approximate number.

Approximate number	3.6	4.9	5.01	10.00	7.540
Minimum original number	3.55				
Maximum original number	3.64				

5 Consolidating and enhancing

3. Perpendicular and parallel

Pupil Textbook page 100 Level A

1. Read these statements and fill in the brackets.

a. The distance between two parallel lines (　　　　).

b. The two opposite edges of the cover of a mathematics book (　　　　) each other, the two adjacent edges (　　　　) each other.

c. Fold a square piece of paper in half twice, as shown in the diagram below. The angle between the two folds is a/an (　　　　) angle.

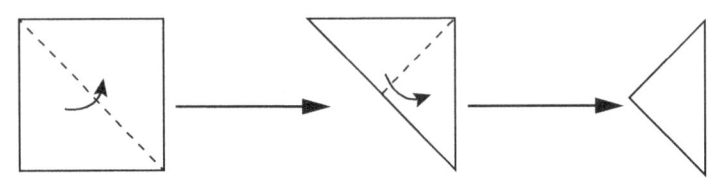

2. Multiple choice – write the letter of the correct answer in the brackets.

a. It is possible to draw (　) vertical line(s) perpendicular to a known horizontal line.

　A. 1　　　　**B.** 2　　　　**C.** infinitely many

b. If two lines are perpendicular to each other, then the angle at their intersection is (　) degrees.

　A. 180　　　**B.** 90　　　**C.** 45

c. If the lines parallel to a known line are drawn through a given point outside the known line, the number of parallel lines will be (　).

　A. 1　　　　**B.** 2　　　　**C.** infinitely many

d. In the diagrams below, there is only one pair of perpendicular lines in diagram (　), and there are three pairs of parallel lines in diagram (　).

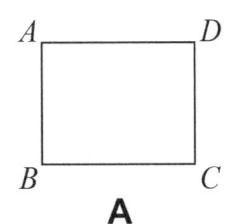

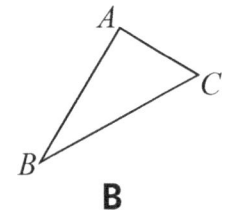

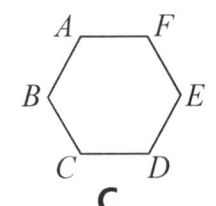

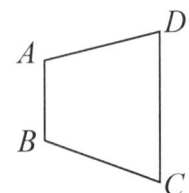

　　A　　　　　B　　　　　C　　　　　D

3. Which lines in this diagram are perpendicular to each other? Which lines are parallel to each other?

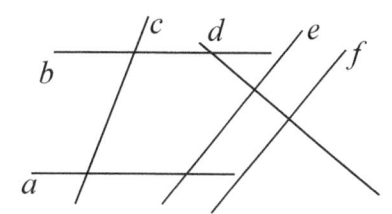

The lines that are perpendicular to each other are: _____

The lines that are parallel to each other are: _____

4. Draw a line through point A that is parallel to line a. Draw a line through point A that is perpendicular to line b.

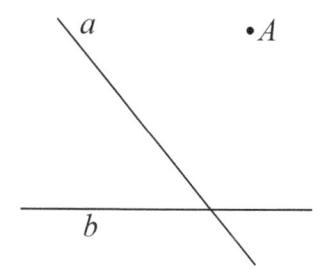

5 Consolidating and enhancing

Level B

1. Look at the cube on the right.

    Which line segments are parallel to line segment *DH*? _____

    Which line segments are perpendicular to line segment *BC*? _____

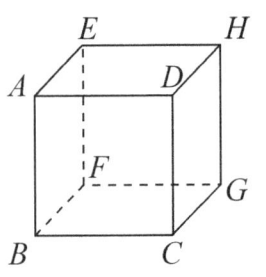

2. a. Draw lines parallel to the diagonals of this rectangle so that the parallel lines pass through the four vertices of the rectangle.

    b. Measure the distance from point *D* to line segment *AC*: _____ cm. (Give your answer to 1 decimal place.)

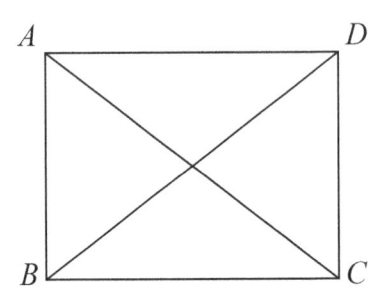